# 5th International Conference on Science, Technology, and Interdisciplinary Research (IC-STAR 2019)



## Interdisciplinary Research for Knowledge Advances for Nation Building

IOP Conference Series: Materials Science and Engineering
Volume 857

Bandar Lampung, Indonesia
23 - 25 September 2019

**Editors:**

**Khairudin**
**Denni Kurniawan**
**Joni Agustian**
**Ahmad Kafrawi Nasution**
**Hery Dian Septama**

ISBN: 978-1-7138-2807-5
ISSN: 1757-8981

**Printed from e-media with permission by:**

Curran Associates, Inc.
57 Morehouse Lane
Red Hook, NY 12571

CURRAN ASSOCIATES INC. proceedings.com

**Some format issues inherent in the e-media version may also appear in this print version.**



Institute of Physics
Dirac House, Temple Back
Bristol BS1 6BE UK

Phone: 44 1 17 929 7481
Fax: 44 1 17 920 0979

techtracking@iop.org

**Additional copies of this publication are available from:**

Curran Associates, Inc.
57 Morehouse Lane
Red Hook, NY 12571 USA
Phone: 845-758-0400
Fax: 845-758-2633
Email: curran@proceedings.com
Web: www.proceedings.com

# TABLE OF CONTENTS

IOP Conf. Series: Materials Science and Engineering **857** (2020) 011001 doi:10.1088/1757-899X/857/1/011001

# PREFACE

The 5th International Conference on Science, Technology, and Interdisciplinary Research (IC-STAR) is organized by Faculty of Engineering, Universitas Lampung, has been held on 23-25 September, 2019 in Bandar Lampung, INDONESIA. IC-STAR 2019 highlighting on the current growth in engineering, science and interdisciplinary research. IC-STAR 2019 theme is "Interdisciplinary Research for Knowledge Advances for Nation Building".

IC-STAR accomplished as a forum to exchange and to discuss information in the most advanced scientific fields related to the research and development in the field of science, engineering and technology, and their inter-discipline subjects among participants from the world's academia, industries, and governments. The Conference also provides a platform for collaborations among professional societies and to enhance technical exchanges among participants.

We are committed on increasing the quality of the papers through a fruitful peer-review based on the originality, impact, novelty, and the relevance to the area of the conference. The published papers were selected from IC-STAR conference 2019 submitted and presented paper.

The committee would like to thank you all distinguished keynote speakers, authors, and reviewers for their contribution that have made this conference a forum of exchange, enlightening and network. Also, for all steering committee and organizing committee member and staff that plan and make the conference possible. We do hope that you would find this publication as an enjoyable and useful guide and reference.

**Editorial board:**

Khairudin, Faculty of Engineering, Universitas Lampung, Indonesia

Denni Kurniawan, Faculty of Engineering, Universiti Teknologi Brunei, Brunei Darussalam

Joni Agustian, Faculty of Engineering, Universitas Lampung, Indonesia

Ahmad Kafrawi Nasution, Faculty Engineering, Universitas Muhammadiyah Riau, Indonesia

Hery Dian Septama, Faculty of Engineering, Universitas Lampung, Indonesia

IOP Conf. Series: Materials Science and Engineering **857** (2020) 011002 doi:10.1088/1757-899X/857/1/011002

# Peer review statement

All papers published in this volume of *IOP Conference Series: Materials Science and Engineering* have been peer reviewed through processes administered by the proceedings Editors. Reviews were conducted by expert referees to the professional and scientific standards expected of a proceedings journal published by IOP Publishing.

# Synthesis of chitosan-$SiO_2$ composite for adsorption methyl dyes from solution

**F Riyanti[1], P L Hariani[1*], Fatma[1], N Yuliasari[1], M Said[1], and T Ramadiati[1]**

[1] Department of Chemistry, Faculty Mathematics and Natural Sciences, Sriwijaya University, Jalan Palembang-Prabumulih Km 32, Ogan Ilir 30662, Indonesia

[*]Email: puji_lukitowati@mipa.unsri.ac.id

**Abstract.** Synthesis of chitosan-$SiO_2$ composite has been done using the precipitation method. The chitosan-$SiO_2$ composite was applied to removal methyl red dyes from the solution. The chitosan-SiO2 composite was characterized using Fourier Transform Infrared Spectroscopy (FT-IR), X-ray Diffraction (XRD), and Scanning Electron Microscopy/Energy Dispersive X-Ray Spectroscopy (SEM-EDS). Adsorption by batch method with variables including contact time, weight of adsorbent and initial concentration. The FT-IR spectrum of composite showed the -OH, C-H, C-N, N-H, C-O (ether), C-H asymmetric, C-H symmetric, Si-O-Si, Si-OH and Si-O group. XRD spectra of chitosan-$SiO_2$ composites showed a widening peak at $2\theta = 19.95$ and amorphous. From SEM-EDS spectra show the morphology of chitosan-$SiO_2$ composite was rough, porous and inhomogeneous with constituent elements C, N, O, Na and Si. The chitosan-$SiO_2$ composite has a pHpzc value at pH 7.9. Adsorption data of chitosan-$SiO_2$ composite for adsorption methyl red dyes followed Langmuir isotherm with maximum adsorption capacity 20.876 mg/g. Adsorption methyl red dye onto chitosan-$SiO_2$ composites according to pseudo-second order.

**Keywords**: Chitosan, $SiO_2$, composite, adsorption, methyl red dye

## 1. Introduction

Disposal of dyes into the environment, reducing aesthetics, serious damage to biological organisms, human health and ecological [1-2]. The dyes used in the industry generally contain complex aromatic structures and are difficult to degrade such as benzene, toluene, xylene [3]. Industries that produce waste containing dyes such as textiles, printing, paint, plastic, cosmetics, paper, and food. The presence of dyes in the environment causes several problems, including reduced ability to re-oxygenate water, inhibit photosynthetic activity, cause an acute and toxic effect on humans [2]. Synthetic dyes have direct and indirect effects that are related to cancer, irritation, tumors, mutations, and allergies [4].

Various methods can be used to reduce waste containing dyes, for example, biological treatment, electrochemical, oxidation, photocatalytic degradation, ion exchange, coagulation, adsorption, and others. Some literature reports that the adsorption method is an effective, rapid, and inexpensive method for reducing dyes [5-6]. Several types of adsorbents from natural and synthetic materials can be used for the adsorption process. Chitosan is an ideal adsorbent and often used as an adsorbent, including biodegradable, non-toxic, biocompatibility, and bioactivity [7-8]. Chitosan can absorb heavy metal ions and dyes [5-7]. Chitosan has a molecular formula poly-β-(1→4)-2-amino-2-deoxy-D-glucose. The functional groups of -OH and $-NH_2$ in chitosan is a role group in the adsorption process [9]. However, chitosan has limitations in its use, including dissolving in acid solution, low in chemical resistance, mechanical strength, and surface area [7].

Modifications of chitosan are to be required to increase adsorption capacity. Some researchers have developed several modifications of chitosan including fiber-chitosan [10], bentonite-chitosan [11], chitosan-montmorillonite [12], magnetic-chitosan [13], and cellulose-chitosan [7]. In this study, chitosan was modified with $SiO_2$ where $SiO_2$ coating on the surface of chitosan. $SiO_2$ is a material that has stability in acidic solutions, thermal stability, and resistance to microbes [14]. Chitosan-$SiO_2$ composites were applied to removal methyl red dyes from aqueous solution. Methyl red is an anionic dye with the molecular formula of $(CH_3)_2NC_6H_4N{=}NC_6H_4CO_2H$, and the molecular weight 269.30 g mol [15]. These dyes can irritate the eyes, skin, and respiratory tract when inhaled [16]. Besides that, methyl red dyes are considered mutagenic for living organisms in anaerobic conditions [17]. Therefore, reducing dyes in wastewater is very important before being discharged into the environment.

## 2. Material and Methods

### *2.1. Materials*

The chemicals used include chitosan from CV Bio Chitosan, Indonesia (95.2 % deacetylation degree), $SiO_2$, NaOH, $C_2H_5OH$, $NaNO_3$, $CH_3COOH$ and HCl from Merck, methyl red dyes from Sigma Aldrich.

### *2.2. Synthesis chitosan-$SiO_2$ composites*

About 5 g of chitosan was dissolved in 100 mL of 1 M $CH_3COOH$ and stirred using a magnetic stirrer at 150 rpm for 2 hours. Then, 2.5 g of $SiO_2$ was added to the chitosan solution and stirred for 1 hour until homogeneous. The mixture is stored for 8 hours. NaOH solution (15% in 95% ethanol) as much as 50 mL was added to the mixture using a dropper pipette while stirring using a magnetic stirrer. The mixture stored for 24 hours for composite formation. The composites were separated from the solution and washed with distilled water until pH ± 7. Then, composite dried in an oven at a temperature of 50°C. The composite is crushed and sieved with a size of 100 mesh. The composite of chitosan-$SiO_2$ powder was characterized using FT-IR (8201PC Shimadzu), XRD (Miniflex 600), and SEM-EDS (Jeol JSM-6335F).

### *2.3. Determination of $pH_{PZC}$*

Determination of $pH_{PZC}$ obtained by adding 0.2 g of the composite to 20 mL of the $KNO_3$ solution. The pH of the solution was adjusted from 2-11 by adding 0.1 M of HCl or NaOH solution. By the shaker, the mixture was stirred for 24 hours. Then the final pH is determined using a pH meter.

### *2.4. Adsorption experiment*

The adsorption process was carried out by a batch experiment using a thermostatic shaker. 50 mL of 50 mg/L methyl red dye solution was added 0.05 g of the chitosan-$SiO_2$ composite. Effect of contact time obtained with variations in contact times of 20, 40, 60, 80, 100, and 120 minutes, stirred using a shaker at 150 rpm at room temperature. The concentration of filtrate was analyzed using a UV-Vis spectrophotometer at wavelength 524 nm (Genesys 10S). The effect of the weight of adsorbent was carried out by the same procedure as the variation in weight of 0.05, 0.10, 0.15, 0.20, 0.25, and 0.3 g. Variations in concentration were observed at initial concentrations of 25, 50, 75, 100, 125, and 150 mg/L. The adsorption ability chitosan-$SiO_2$ composite compared with chitosan.

## 3. Result and Discussion

### *3.1. Characteristic of chitosan-$SiO_2$ composite*

Formation of silica on chitosan can occur through hydrogen bonds or Van der Waals. Chitosan has many hydroxyl groups which can form hydrogen bonds with silanol groups as a result of hydrolysis. Another bonding mechanism between silica and chitosan by electrostatic attraction of chitosan undergoes protonation in an amino group and a hydroxyl group dissociating of silica in solution [18]. The XRD spectra of chitosan, $SiO_2$, and chitosan-$SiO_2$ composites are shown in Figure 1. The XRD spectra of chitosan-$SiO_2$ composites appear diffracted widened at $2\theta = 5$ to $55^{o}$, minor diffraction is detected at $10.32^{o}$ and the highest peak at $20.72^{o}$.

These composite spectra appear to be a combination of $SiO_2$ and chitosan diffraction. This study is like other studies that chitosan-$SiO_2$ composites synthesized using ultrasonic batch have diffraction peaks at 10.4 and $23.6^{o}$ [19]. A widening diffraction peak indicates amorphous silica. The highest peak at $2\theta = 22.52^{o}$. According to JCPDS No. 39-1425 that the silica has a peak at $2\theta = 22.003^{o}$. Amorphous silica is the most reactive and suitable to be applied as an adsorbent [20]. Chitosan also has a widening peak with a sharp peak at $2\theta = 20.321^{o}$ (JCPDS 39-1834). In this study obtained at $20.23^{o}$.

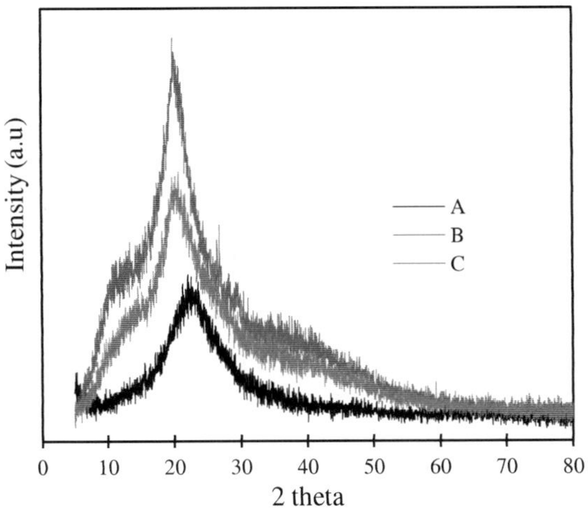


**Figure 1**. XRD pattern of (A)$SiO_2$ (B)chitosan (C)chitosan-$SiO_2$ composite.

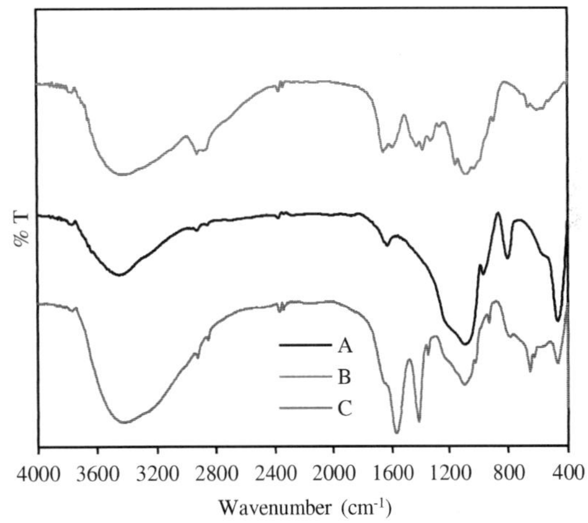


**Figure 2**. FT-IR spectra of (A)$SiO_2$ (B)chitosan (C)chitosan-$SiO_2$ composite.

FT-IR analysis was carried out in the wavenumber range 400-4000 $cm^{-1}$. FT-IR spectra of $SiO_2$, chitosan, and chitosan-$SiO_2$ composite displayed in Figure 2. Chitosan has a wide peak at 3421.51 $cm^{-1}$ (overlap between O-H and N-H groups). The peak at 2877.79 $cm^{-1}$ is C-H stretch. Amide groups can be observed at wavenumbers 1656.85, 1595.13 and 1323.17 $cm^{-1}$, respectively. C-O stretch is detected at 1083.99 $cm^{-1}$. The characteristics $SiO_2$ peak are the hydroxyl O-H of the silanol group and water that appears at wavenumbers 3450.65 and 1631.78 $cm^{-1}$, Si-O groups at 1095.57, 970.19, 802.39 and 466.71 $cm^{-1}$ which show Si-O stretching, Si-OH stretching, Si-O stretching and Si-O-Si bending. The peak in the composite is a combination of $SiO_2$ and chitosan. The wavenumber at 1413.82 and 1423.47 $cm^{-1}$ shows the presence of C-H groups (asymmetric) on chitosan-$SiO_2$ composite and chitosan which are not present in $SiO_2$. This study shows that composite synthesis has been successful.

Figure 3 shows the morphology of chitosan and chitosan-$SiO_2$ composites. The composite surface appears heterogeneous and dense while the surface of chitosan is more homogeneous and porous. Table 1 indicates the different elements between chitosan and chitosan-$SiO_2$ composites from EDS data. The main components of chitosan and chitosan-$SiO_2$ composite are C, N, and O. In the composite, there is the addition of Si elements from $SiO_2$ (11.76%). The presence of Na on the composite from NaOH for synthesis has not disappeared.

IOP Conf. Series: Materials Science and Engineering **857** (2020) 012001 doi:10.1088/1757-899X/857/1/012001

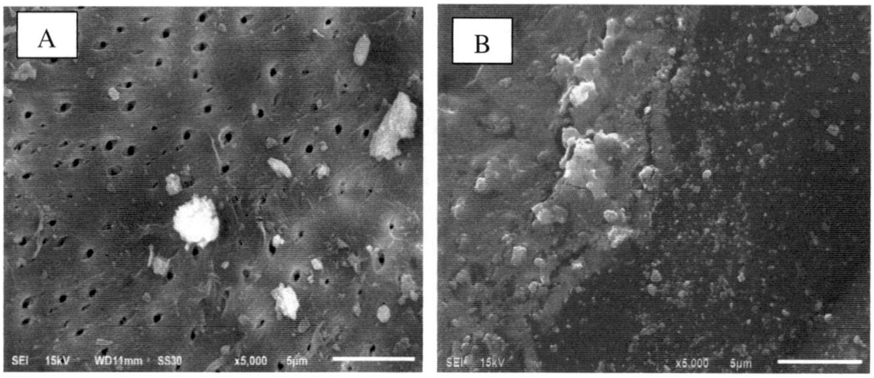


**Figure 3**. SEM image for (A)chitosan, (B)chitosan-$SiO_2$ composite.

**Table 1**. Elements of chitosan and chitosan-$SiO_2$ composites.

| Elements (%) | Adsorbents | |
|---|---|---|
| | Chitosan | Chitosan-$SiO_2$ composite |
| C | 47.41 | 17.98 |
| N | 48.35 | 24.63 |
| O | 2.64 | 41.74 |
| Si | - | 11.76 |
| Na | - | 3.90 |
| Ca | 1.60 | - |

*3.2. $pH_{PZC}$ chitosan-$SiO_2$ composite*

pH is one of the parameters that play a role in the adsorption process. The adsorption capacity is influenced by the pH of the solution, correspondent to the electrostatic pull between the adsorbate and the adsorbent. $pH_{PZC}$ chitosan-$SiO_2$ composite was obtained 7.9 while chitosan at pH 7.7. $pH_{PZC}$ of chitosan-$SiO_2$ composite and chitosan is shown in Figure 4. The pH of the solution < $pH_{PZC}$, the adsorbent is positively charged while the pH > $pH_{PZC}$, negatively charged of the adsorbent. Methyl red dye has pKa 5.1. The effectiveness of the adsorption process at the pH of the dye solution (±5.1) where there is an electrostatic attraction between negatively charged methyl red dyes and positively charged chitosan or chitosan-$SiO_2$ composites. Other research, adsorption of methyl red dyes using activated carbon activated from custard apple shell using $K_2CO_3$, the optimum pH is 5 [15].

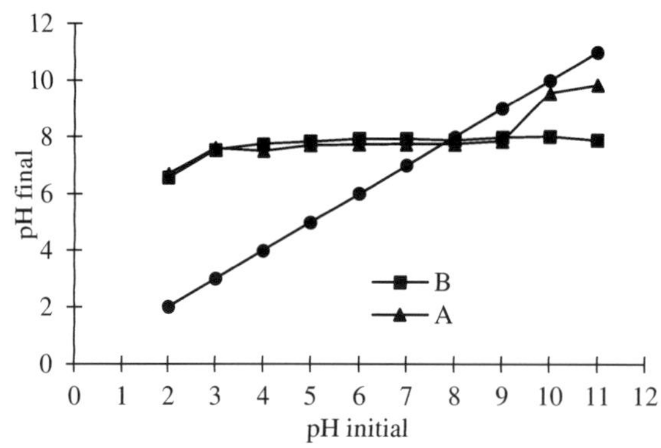


**Figure 4**. $pH_{PZC}$ (A)chitosan and (B)chitosan-$SiO_2$ composite.

### *3.3. Adsorption study*

In this study, three variables were used to study the adsorption process, including the effect of contact time, the weight of adsorbent, and the initial concentration of the dye. The effect of adsorption variables on the removal of dyes adsorbed by chitosan-$SiO_2$ composite and chitosan was illustrated in Figure 5. From Figure 5A, the adsorption equilibrium using chitosan-$SiO_2$ composites is faster than chitosan. Also, the amount of dyes removal by the composite is greater than chitosan. $SiO_2$ coating on the surface of chitosan causes the addition of active groups on the surface of chitosan so that more active sites to absorb dye. The adsorption equilibrium was obtained at 40 minutes using a chitosan-$SiO_2$ composite while chitosan was obtained at 100 minutes.

In Figure 5B, it can be seen that the % removal of methyl red dyes by chitosan, and chitosan-$SiO_2$ composites increase as the number of adsorbents increases. The increase in weight of the adsorbent is proportional to the increase in the availability of active sites. The limited amount of the dye causes the addition of the adsorbent to not affect the removal of dye. The adsorption capacity of the chitosan-$SiO_2$ composite is greater than chitosan, as shown in Figure 5C. The presence of $SiO_2$ in the composite adds to the active site, which can bind to the methyl red dye. Besides that, the addition of silica also improves the mechanical properties of chitosan [21].

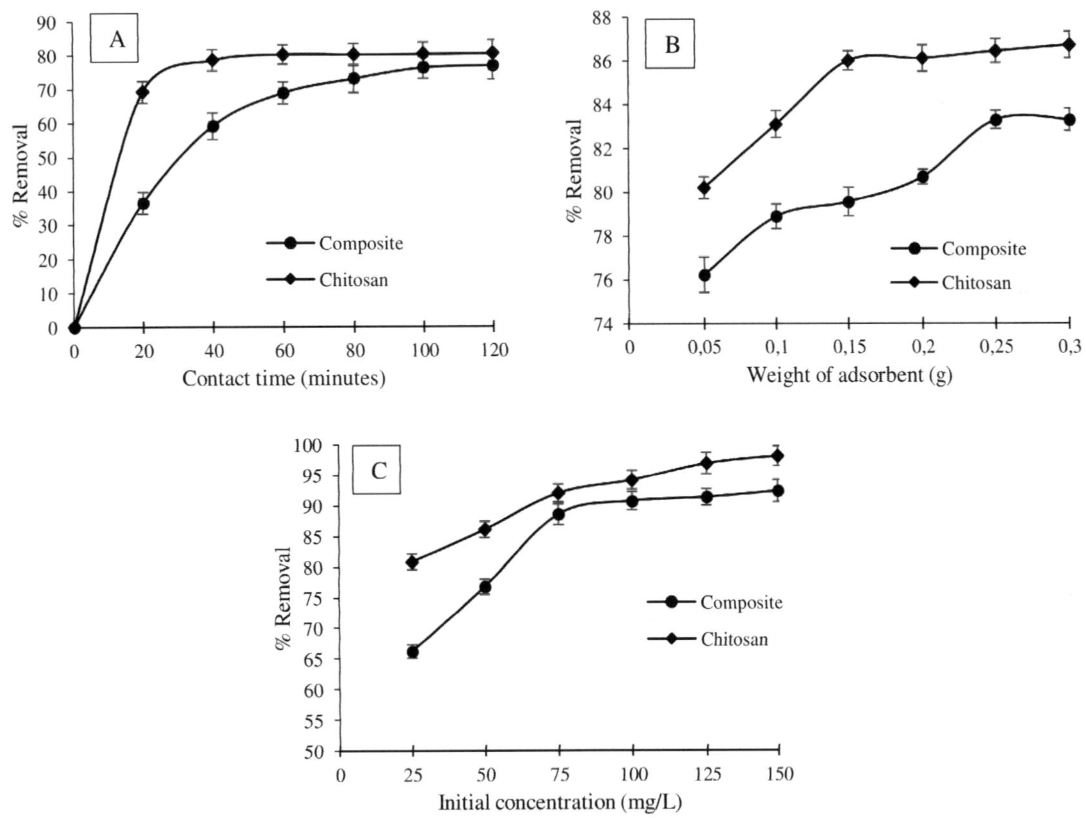


**Figure 5**. Sorption of methyl red dye onto chitosan and chitosan-$SiO_2$ composite (A)effect contact time (B)effect weight of adsorbent and (C)effect of initial concentration.

The adsorption isotherm is important to describe adsorbate and adsorbent interactions in solid-liquid adsorption [10]. In this study, adsorption isotherm used the Langmuir and Freundlich isotherm models. The Langmuir isotherm equation illustrates that the surface of adsorbent a monolayer and is homogeneous, whereas for Freundlich's isotherm [7]. The Langmuir and Freundlich equations are expressed in the following formula:

$$q_e = \frac{q_m K_L C_e}{1 + K_L C_e} \quad (1)$$

$$q_e = K_F C_e^{1/n} \quad (2)$$

Where $q_e$ is the concentration of the dye at the equilibrium onto adsorbent (mg/g), $C_e$ is the concentration of the dye in the solution (mg/L), $q_m$ is adsorption capacity of adsorbent (mg/g), $n$ is the value associated with the intensity of adsorption, $K_L$ and $K_F$ are the Langmuir (L/mg) and Freundlich (mg/g) constant, respectively. The parameters of adsorption isotherm are shown in Table 2. Suitability of isotherm can be identified from the value of the correlation coefficient ($R^2$) [22]. The $R^2$ value of Langmuir and Freundlich isotherm are 0.994 and 0.959, respectively. Therefore, isotherm Langmuir is more suitable for describing the adsorption process of methyl red dye onto chitosan and chitosan-$SiO_2$ composites. The value of $n$ between 2-5 indicates that it is a favourable process.

**Table 2.** The Parameters of isotherm adsorption methyl red dye onto chitosan and composite.

| Langmuir isotherm | | | Freundlich isotherm | | |
|---|---|---|---|---|---|
| Parameters | Chitosan | Composite | Parameters | Chitosan | Composite |
| $R^2$ | 0.994 | 0.991 | $R^2$ | 0.939 | 0.959 |
| $q_m$(mg/g) | 10.234 | 20.876 | $n$ | 2.345 | 3.213 |
| $K_L$ (L/mg) | 1.870 | 1.349 | $K_F$ (mg/g) | 2.410 | 4.782 |

Adsorption kinetics can be used to describe the adsorption mechanism. The adsorption kinetics depends affected by the chemical or physical characteristics of adsorption. Pseudo-first order and pseudo-second order are often used, kinetics models. The first-order, and second-order are the following:

$$\log(q_e\text{-}q) = \log q_e - \frac{k_1}{2.303} t \quad (3)$$

$$\frac{t}{q_e} = \frac{1}{k_2 q_e^2} + \frac{t}{q_e} \quad (4)$$

Where $q_e$ and $q$ are the amount of dyes absorbed at the equilibrium and time (mg/g), $k_1$ and $k_2$ are constant in pseudo-first order and pseudo-second order. Adsorption kinetics data are shown in Table 3. The adsorption of methyl red dye on chitosan and chitosan-$SiO_2$ composites is more appropriately described using the pseudo-second order based on the value of $R^2$. Another study reported that adsorption of methyl red onto banana pseudo-stem fibers and $MnFe_2O_4$-activated carbon composites and followed a pseudo-second order [22-23].

**Table 3**. The Parameters of kinetic adsorption methyl red dye onto chitosan and composite.

| Pseudo-first order | | | Pseudo-second order | | |
|---|---|---|---|---|---|
| Parameters | Chitosan | Composite | Parameters | Chitosan | Composite |
| $R^2$ | 0.893 | 0.903 | $R^2$ | 0.936 | 0.992 |
| $k_1$(min$^{-1}$) | 0.055 | 0.072 | $k_2$ (g/mg.min) | 0.0025 | 0.0031 |
| $q_1$ (mg/g) | 8.982 | 13.421 | $q_2$(mg/g) | 4.112 | 7.819 |

## 4. Conclusions

In this study, chitosan-$SiO_2$ composite by precipitation method was successfully synthesized. The chitosan-$SiO_2$ composite has a greater adsorption capacity to methyl red dyes than chitosan. Isotherm Langmuir is more suitable for describing adsorption methyl red dye onto chitosan-$SiO_2$ composites with adsorption capacity 20.876 mg/g. Kinetic studies show that adsorption was in accordance the pseudo-second order. The adsorbents can be used as an option to removing methyl red dye from the solution.

**Acknowledgments**
This research was funded from the SATEKS Universitas Sriwijaya grant in 2019 (No. 0146.177/UN9/SB3.LP2M.PT/2019).

## References

[1] Prasad A L and Santhi T 2012 *Sustain. Environ. Res.* **22** 2 113
[2] Esquerdo VM, Cadaval TRS, Dotto GL and Pinto L AA 2014 *J. Colloid Interface Sci.* **424** 7
[3] Sumanjit K, Seema R and Rakesh K M 2013 *J. Chem.* **2013** 1
[4] Hariharasutan R, Nageswara R A and Bhaskaran A 2013 *Int. J. Eng. Sci.* **2** 287
[5] Mohammadtaghi V, Mohd R, Babak S, Ahmad Z A, Mahamad H I, Kok B T, Zahra G and Parisa A 2014 *Carbohydr. Polym.* **113** 115
[6] Bhavani K, Roshan A B E, Selvakumar S and Shenbagarathal 2016 *J. Bioremediat. Biodegrad.* **7**
[7] Yixi W, Hao W, Huili P, Zhicun W, Jianning W and Zhiyong L 2018 *Fiber. Polym.* **19** 340
[8] George Z K and Dimitrios N B 2015 *Mar. Drugs.* **13** 312
[9] Jinyun L, Yu C, Tianli H, Mengyin C, Wen Z, Jiawei L and Xiangqian F 2019 *Chemosphere* **214** 738
[10] Rongjun Q, Changmei S, Minghua W, Chunnuan J, Qiang X, Ying Z, Chunhua W, Hou C and Ping Y 2009 *Hydrometallurgy* **100** 65
[11] Hariani P L, Fatma, Fahma R and Hesti R 2015 *J. Manusia dan Lingkungan* **22** 233
[12] Thanut J and Walaikorn N 2016 *Mat. Res.* **19** 1114
[13] Jixiang L, Buqing J, Yi L, Changquan Q, Jiajun H, Guangren Q, Wenshan G and Huu H N 2017 *J. Clean. Prod.* **158** 51
[14] Tetyana MB, Levgen V P, Valentin A T, Elina S Y and Dorota K 2015 *Nanoscale Res. Lett.* **10** 1
[15] Equbal A K, Shahjahan and Tabrez A K 2018 *J. Mol. Liq.* **249** 1195
[16] Badr Y, Abd El Wahid M G and Mahmoud M A 2008 *J. Hazard Mater.* **154** 245
[17] Kim T C, George E F and Andrew W 1981 *Appl. Environ. Microbiol.* **42** 641
[18] Tetyana B, Valentin T and Elina Y 2014 *Mater. Sci-Medzg.* **20** 177
[19] Sayyid M R and Amir H 2017 *Sustain. Chem. Eng.* **5** 10379
[20] Chandrasekhar S, Satyanarayana K G, Pramada P N and Raghavan P 2003 *J. Mater. Sci.* **38** 3159
[21] Mahatmanti F W, Nuryono and Narsito 2014 *Indo. J. Chem.* **14** 131
[22] Haris M R H M and Kathiresan S 2009 *Am. J. Appl. Sci.* **6** 1690
[23] Riyanti F, Hariani PL, Purwaningrum W, Elfita, Damarril SS and Amelia A 2018 *Molekul* **13** 123

# Antioxidant Activities Bioactive Compound of Ethyl Acetate Extracts from rose myrtle Leaves (*Rhodomyrtus tomentosa* (Ait.) Hassk.)

**Salni[1*], Hanifa Marisa[1]and Lili Arista Repi[1]**

[1] Department of Biology, Faculty Mathematics and Natural Sciences, Sriwijaya University, Ogan Ilir, Indonesia

*Email: salnibasir@unsri.ac.id

**Abstract.** One of the plants traditionally used as medicine is rose myrtle (*Rhodomyrtus tomentosa* (Ait.) Hassk.). Rose myrtle has the potential as an antioxidant because it has secondary metabolites including flavonoids, phenols, terpenoids, tannins, saponins, and steroids. The purpose of this study was to determine the fraction that has antioxidant activity and obtain antioxidant compounds from ethyl acetate extract of rose myrtle leaves. The research methods included extraction by Soxhletasi, fractionation by vacuum liquid chromatography, purification of compounds with gravity columns, and antioxidant activity test using the DPPH method. The results obtained 32% yield of ethyl acetate extract which was then followed by fractionation. The fractionation results obtained 13 fractions. The fractions that have antioxidant activity namely fractions 5-6, 7, and 8. The fraction was continued to purification and obtained by compound E.1, compound E.2 from the essential oil, and compound E3 allegedly rhodomyrtone. The bioactive compound determined by $IC_{50}$ value obtained that E1, E2 and E3 that have $IC_{50}$ value of 218, 207 and 90 μg/mL, respectively. The bioactive compounds of ethyl acetate extract have strong antioxidant activity in E3 and very weak in E1 and E2 compounds.

**Keywords:** Antioxidant, *Rhodomyrtus tomentosa, e*ssential oils, free radical, DPPH

## 1. Introduction

Free radical compounds are one of the factors in DNA damage. If the damage is not too severe, the DNA system can repair it. However, if damage has occurred in various places, cannot be repaired by a DNA system so that cell division will be disrupted. This damage causes abnormal changes in the gene and can cause cancer [1]. Antioxidants are compounds that can bind free radicals and molecules that are very reactive. Antioxidants will give hydrogen atoms that are oxidant so that cell damage in the body can be inhibited. This compound has a small molecular weight and can inactivate the development of oxidation reactions by preventing radical formation [2].

Natural antioxidants derived from plants are safer than synthetic antioxidants such as propyl gallate, citric acid, butylated hydroxyanisole, and butylated hydroxytoluene which can cause hypoxic and carcinogenic [3]. Synthetic antioxidants such as BHA (butylated hydroxy aniline) and BHT (butylated hydroxy toluene) have great side effects including causing liver damage.

On the other hand, nature provides an effective and relatively safe source of antioxidants such as flavonoids, vitamin C, beta carotene and others. This encourages more exploration of natural materials as a source of antioxidants [4].

One of the plants in Indonesia that is traditionally used as medicine is rose myrtle plants (*Rhodomyrtus tomentosa* (Ait.) Hassk.). According to other research [3] that rose myrtle leaf extract has strong antioxidant activity in vitro and in vivo. Therefore, rose myrtle extract can be used as an antioxidant supplement in food products to prevent food oxidation and also as an alternative antioxidant therapy in the pharmaceutical industry to prevent free radical damage. Further research is needed to evaluate the toxicity and antioxidant activity of each phytochemical compounds to develop a complete supplement/antioxidant therapy. The ethyl acetate fraction from rose myrtle leaf extract has antioxidant activity with $IC_{50}$ value of 70.12 μg/mL [5]. The important compound in antioxidant activity in the fraction is phenols and flavonoids. In this study, was to isolate antioxidant compounds from ethyl acetate fraction of rose myrtle leaves.

## 2. Material and methods

### *2.1. Materials*
The material used is *Rhodomyrtus tomentosa* leaves obtained from the flat land of West Sumatra Sungayang district. The chemicals used include ethyl acetate, methanol, n-hexane, DMSO, DPPH *(1,1-Diphenyl-2-pikrihidrazil)*, Silica gel 60, and silica gel 60 F254 from Merck.

### *2.2. Extraction*
Extraction using Soxhlet with ethyl acetate. 150 g of simplicia powder wrapped in filter paper put into Soxhlet, added with ethyl acetate 1 L. The liquid ethyl acetate extract obtained was evaporated with a rotary evaporator at a temperature of 80°C until thick extracts were obtained.

### *2.3. Fractionation*
Fractionation was carried out using the VLC (Vacuum Liquid Chromatography) method, with a stationary phase in the form of silica gel and various stages of the mobile phase of solvents. Silica gel was inserted into the VLC column with 5 cm silica height. The solvents used were n-hexane, ethyl acetate, and methanol with a n-hexane ratio (%) of 100:0; 95:5; 90:10; 85:15; 80:20; 75:25; 70:30; 65:45; 60:40; 55:55; 50:50, 100% ethyl acetate, and 100% methanol as much as 200 mL.

### *2.4. Antioxidant test of fraction*
The fraction of the ethyl acetate extract was bottled on the TLC plate using a capillary pipette, then inserted into the chamber containing n-hexane: ethyl acetate (8:2) eluent. Furthermore, the eluent was allowed to propagate to the TLC plate boundary. TLC was left in the air to dry. The dry TLC was sprayed using DPPH 0.5 mM. Color changes in TLC were observed. Spots from the fraction that have antioxidant activity will turn yellow with a purple background. Spots that appear then calculated the RF (Retardation Factor) value.

### *2.5. Compounds purification*
Purification of compounds from the active fraction was carried out by gravity column chromatography with the stationary phase of silica gel, while the mobile phase was a mixture of n-hexane and ethyl acetate and methanol. The polarity was increased gradually. Elution was carried out by eluent solvent mobile phase which corresponds to the elution rate of 17 drops per minute, with the volume of the fraction collected being 10 mL. The pure compounds obtained were then tested for antioxidant activity and determined IC50 values.

### *2.6. Antioxidant activity*

Pure compounds tested antioxidant activity using the DPPH method. DPPH solution is made with a concentration of 0.5 mM, as much as 1.98 mg DPPH is dissolved in 100 mL of methanol. 1 mg pure compound dissolved in 1 mL DMSO (1000 ug/mL). The solution was diluted with concentrations (500, 250, 125, 62.5, 31.25, and 15,625) µg/mL. 0.2 mL of each concentration was added 3.8 mL of DPPH 0.5 mM solution, the mixture was homogenized and left for 30 minutes in a dark place. The absorbance value was determined by UV-VIS spectrophotometer at λ max 517 nm. Positive control is used ascorbic acid. Determination of antioxidant activity is determined calculating the formula (1) as follows :

$$\%\ \text{Radical Inhibition DPPH} = \frac{\text{absorbance control} - \text{absorbance sample}}{\text{absorbance control}}\ \text{x}\ 100\ \% \tag{1}$$

## 3. Results and Discussions

### *3.1. Extraction and fractionation*

The extraction of 150 g of simplicia was obtained by 48 g extract with a yield of 32%. The extract was fractionated by the VLC (Vacuum Liquid Chromatography) method using a solvent ratio that varied in degree of polarity. The fractionation results obtained 13 fractions, the results of the antioxidant activity test with DPPH obtained 3 active fractions namely 5-6, 7 and 8 fractions characterized by yellow spots on the VLC plate. The RF value of antioxidant compounds in 2-5 fractions was 0.51; fraction 6 was 5.54 while fraction 8 was 0.33 with n-hexane eluent: ethylacetate (8: 2).

The RF value of the antioxidant compounds obtained has the same value as the research of [5} The value of Rf spotting that has antioxidant activity in the ethylacetate extract of rose myrtle leaves is 0.54; 0.51; 0.33 with n-hexane eluent: ethylacetate (8: 2). Fractions containing antioxidant compounds, 5-6, 7 and 8 fractions were purified using the gravity column chromatography method.

### *3.2. Antioxidant activity of active compounds*

The purification of antioxidant compounds found 3 bioactive compounds, namely compounds E1, E2, and E3. The antioxidant activity of compounds E1, E2, E3, and ascorbic acid can be seen in Figures 1 and 2. In Figure 2 there is a discoloration of DPPH solution with the addition of E1 (A) compounds, E2 (B) compounds, E2 (C) compounds, and ascorbic acid (D) with concentrations (500 µg/mL; 250 µg/mL; 125 µg/mL; 62.5 µg/mL; 31.25 µg/mL; 15.125 µg/mL).

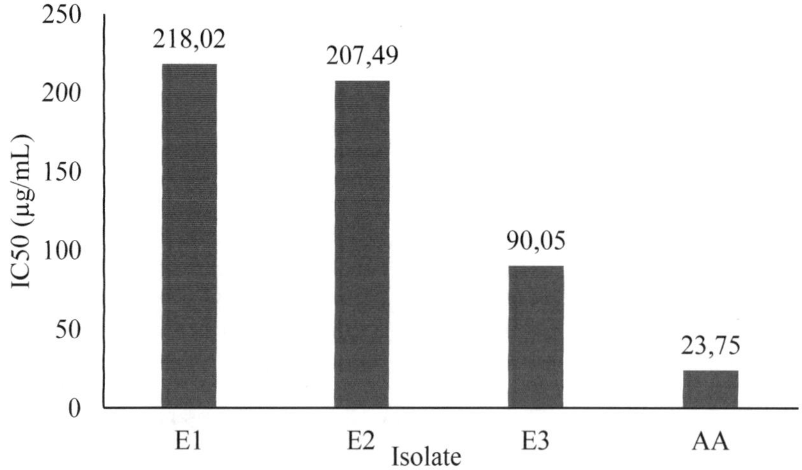


**Figure 1.** The antioxidant activity of compounds E1, E2, E3, and ascorbic acid (AA).

Figures 1 and 2 obtained that E1 compound have an $IC_{50}$ value of 218 ug/mL so that it is categorized as having very weak antioxidant activity. E2 compound had an $IC_{50}$ value of 207.49 μg/mL, it was categorized as having very weak antioxidant activity. E3 compound has an $IC_{50}$ value of 90.05 ug/mL classified as a strong antioxidant activity. The $IC_{50}$ value of the three isolates has $IC_{50}$ under ascorbic acid. Ascorbic acid has very strong antioxidant activity with $IC_{50}$ 23.75 ug/mL. Comparing to others [6], it shows that antioxidant activity can be classified into several categories: under 50 ug/mL is very strong, between 51-100 ug/mL is strong, 101-150 ug/mL, between 151-200 ug/mL is strong and greater than 200 ug/mL is very weak.

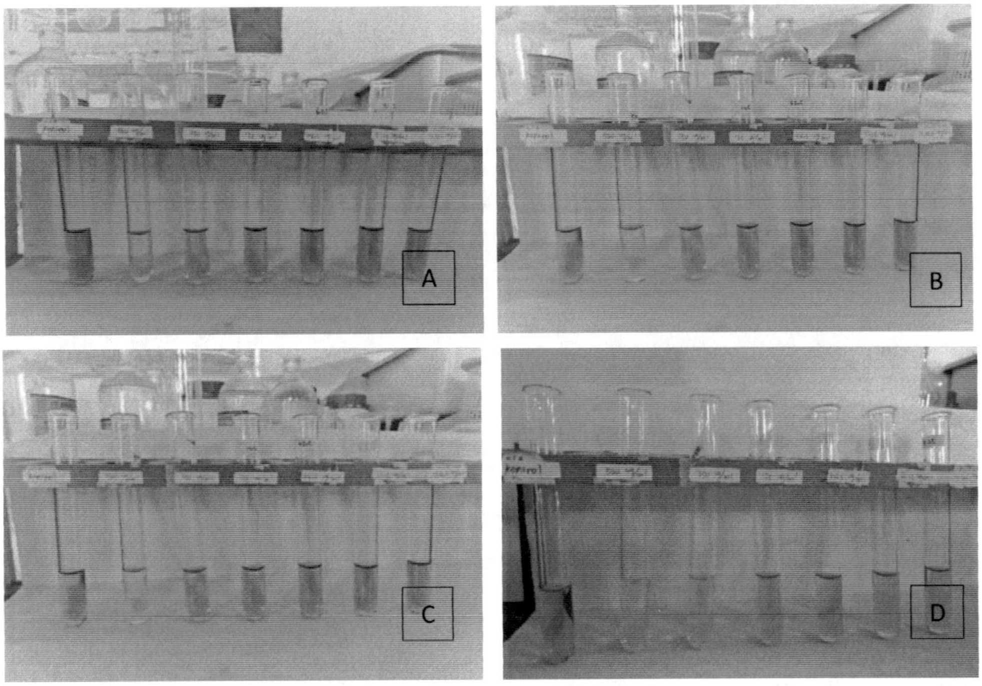


**Figure 2.** Change the color of the solution (A) E1, (B) E2, (C) E3 compounds and (D) ascorbic acid.

The characterization of E1 and E2 bioactive compound with GC-MS are known to be essential oils. The chromatogram of E3 compouns is thought to be rhodomyrtone. Some types of plants have essential oils that have antioxidant activities, including: antioxidant activity of 423 essential oils from 48 plant species obtained 73 essential oils showing $IC_{50}$ concentration of 1.25 mg/mL. $IC_{50}$ of the 73 most active oils ranges from 4 to 2000 μg/mL. Essential oils which have $IC_{50}$ of fewer than 300 μg/mL of 20 samples. Essential oils from the Lamiaceae and Myrtaceae families are the most effective antioxidants. Thymol and carvacrol are the main compounds in most essential oils of the Lamiaceae and eugenol families and terpenes are the main compounds in all essential oils of the Myrtaceae family [7]. Essential oil from *Origanum vulgare* L. with main components as thymol (37.13%), gamma-terpinene (9.67%), carvacrol (9.57%), carvacrol methyl ether (6.88), cis-alphababolabolene ( 6.80%)), eucalyptol (3.82%), p-cymene (3.58%) and elemol (2.04%) [8].

Antioxidant activity of essential oils of E1 and E2 compound from plants rose myrtle is still under the antioxidant activity of red betel essential oil (*Piper crocatum* Ruiz & Pav.) With $IC_{50}$ value 136,947 ug/mL. Essential oils of red betel leaves were analyzed by GC-MS to determine the constituent components of the oil. The chemical components of red betel leaf essential oils which have the largest percentage based on area percent are sabinene (33.35%), β-mycenae (20.18%), β-caryophyllene (7.07%), linalol (5.41%) and germakren (4.96%). The results of GC-MS analysis of red betel leaf essential oil showed that the main constituents of essential oils were monoterpene and sesquiterpene groups [9]

### *3.3. Characterization of E1 isolates*

Isolate E1 in the form of a yellow paste is thought to be an essential oil. Chromatogram of E1 compound is shown in Figure 3. The E1 isolate obtained that 31 peaks spectra which showed there were 31 types of compounds contained in the essential oil. Isolate E1 contained six main compounds namely peak 8.08 semimyrtucommulone compound (15.63), peak 8.10 semimyrtucommulone (6.97), peak 12.91 caryophyllene (13.12), peak 13.17 menthol (29.49), peaks of 13.20 menthol (11.31) and peaks of 13.38 cubenol (10.63), menthol compounds were the compounds with the highest E1 isolates of 29.49%.

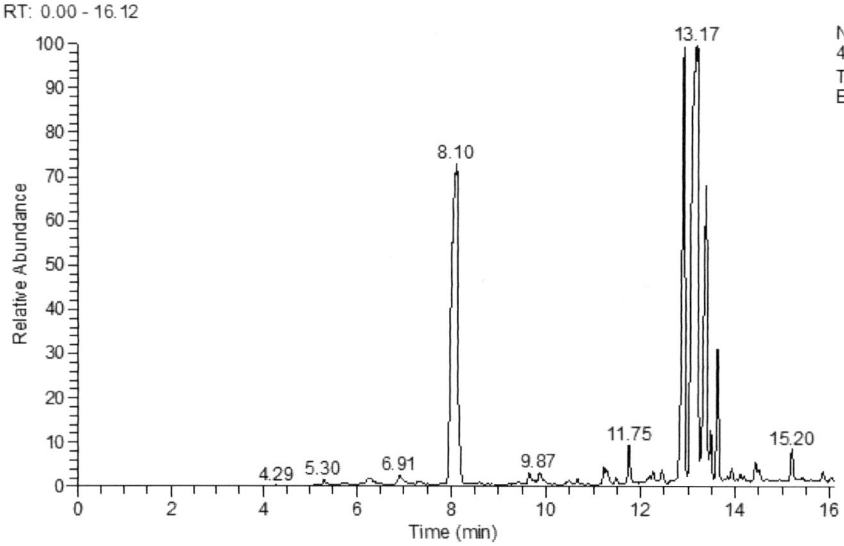


**Figure 3.** Chromatogram of E1 compound.

### *3.4. Characterization of E2 isolate*

E2 isolate in the form of a reddish colored paste. E2 isolate, is thought to be an essential oil. The determined E2 isolate by GC-MS shown in Figure 4. The E2 isolate contained 30 compounds of five major compounds namely Peak 8.08 semimyrtucommulone (15.63), peak 8.10 semimyrtucommulone (6.97), peak 12.91 caryophyllene (13.12), peak 13, 17 menthol (29.49), peak of 13.20 menthol (11.31), peak of 13.38 cubenol (10.63), and menthol compound is the most abundant E1 isolate is 29.49%

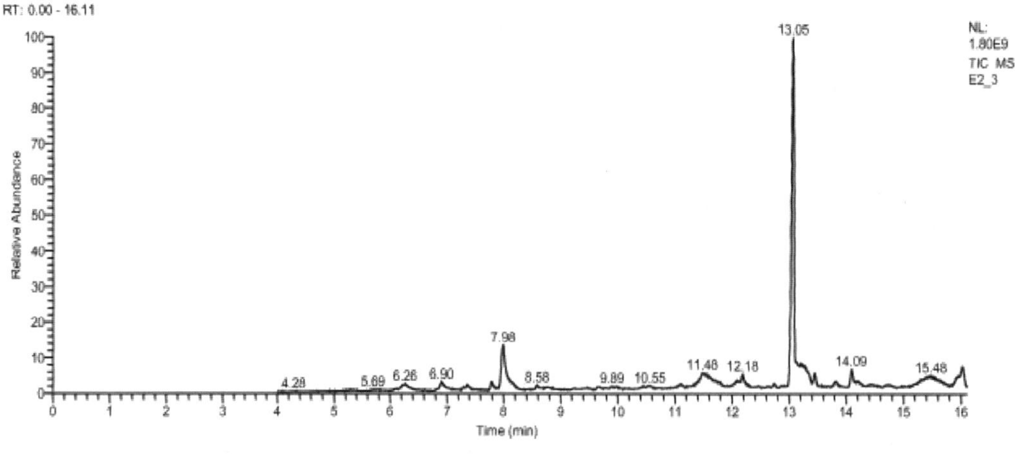


**Figure** 4. Chromatogram of E2 compound.

### *3.5. Characterization of E3 isolate*

E3 isolate in the form of white crystal, chromatogram E3 isolate using thin layer chromatography (TLC) method with mobile phase n-hexane: ethyl acetate (8: 2) obtained rf value of 0.24, after spraying 10% $H_2SO_4$ there are yellowish-orange spots which indicate that these antioxidant compounds belong to the class of phenol compounds. The Rf value and the color of the isolate E3 are the same as the rhodomyrtone compound [10, 11]. Isolate E3 is thought to be the Rhodomyrtone compound, Rhodomyertone including the phenol group, phenol compounds have antioxidant activity because they can bind to free radicals. According to Sari et al. [12], phenol compounds play a significant role in antioxidant activity, the greater the content of phenol groups, the greater their antioxidant activity. This is due to the presence of a hydroxyl group in the phenol group that is capable of capturing free radicals.

## 4. Conclusions

The fractionation of ethylacetate extract produces 13 fractions, which fractions to 5-6, 7, and 8 have antioxidant activity characterized by yellow spots with a purple background. The fractions of 5-6 obtained by E1 compound have $IC_{50}$ value of 218 μg/mL with very weak antioxidant activity, from fraction 7 obtained E2 compound which has $IC_{50}$ value of 207 μg/mL is very weak antioxidant activity. The fraction 8 obtained by the antioxidant E3 compound which has an $IC_{50}$ value of 90 μg/mL with strong antioxidant activity. The $IC_{50}$ value of 3 compound has $IC_{50}$ value under ascorbic acid. The E1 and E2 isolates are essential oils, while the E3 isolates are thought rhodomyrtone compounds.

## Acknowledgment

This research was funded from the Kompetitif Universitas Sriwjaya grant in 2019

## References

[1] Khaira K 2010 Menangkal Radikal Bebas dengan Antioksidan *J. Saintek.* **2** 183-187.
[2] Winarsi H 2007 *Antioksidan Alami dan Radikal* Bebas (Yogyakarta: Kanisius)
[3] Lavanya G, Voravuthikunchai S P and Towatana N H 2012 Acetone Extract from Rhodomyrtus tomentosa : A Potent Natural Antioxidant. *J Evidence-Based Complementary and Alternative Medicine.* **2012** 535479
[4] Bahriul P, Rahman N and Diah AWM 2014 Uji Aktivitas Antioksidan Ekstrak Daun Salam (*Syzygium polyanthum*) dengan menggunakan 1,1-Difenil-2-Pikrilhidrazil. *J Akademika Kimia.* **3** 143-149.
[5] Suryadinata W, Sakti ERE and Kodir RA 2016 Telaah Fitokimia Senyawa Antioksidan dari Ekstrak dan Fraksi Daun rose myrtle (Rhodomyrtus tomentosa (Aiton) Hassk) *Proc. Farmasi (Bandung)* vol 2 (Unisba) p 663-669.
[6] Molyneux P 2004 The use of the stable free radical diphenylpicrylhydrazyl (DPPH) for estimating antioxidant activity Songklanakarin *J. Sci. Technol.* **26** 211-219.
[7] Defaveri A C A, Sato A, Borré L B, Aguiar D L M, Gil R A S S, Arruda R C O and Riehl C A S 2011 *Eugenia neonitida* Sobral and *Eugenia rotundifolia* Casar. (Myrtaceae) Essential Oils: Composition, Seasonality Influence, Antioxidant Activity and Leaf Histochemistry *J. Braz. Chem. Soc.* **22** 1531-1538.
[8] Vazirian, Mohammadi M, Farzaei M H, Amin G, Amanzadeh Y 2015 Chemical composition and antioxidant activity of *Origanum vulgare* subsp.vulgare essential oil from Iran Research. *J of Pharmacognosy.* **2** 41-46
[9] Widayani A, Cahyono A, and Harjono 2018 Isolasi dan Uji Antioksidan Minyak Atsiri Daun Sirih Merah (Piper crocatum Ruiz & Pav.) pada Minyak Goreng Curah. Indo. *J. Chem. Sci.* **7** 214
[10] Salni 2003 Karakterisasi dan Uji Aktivitas Topikal Senyawa Antibakteri dari Daun Karamunting (*Rhodormyrtus tomentosa* (Ait.) Hassk) Disertasi ITB Bandung.

[11] Dachrianus, Salni, Sargent M V, Skelton B W, Soediro I, Sutisna M, White A H and Yulinah E 2002 Rhodomyrtone, an antibiotic from *Rhodomyrtus tomentosa*. *J of Chemistry*. **55** 229-232.
[12] Sari E, Hilma R and Cendrakasih A 2018 Penentuan Aktivitas Antioksidan Ekstrak Biji Dan Daging Buah rose myrtle (*Rhodomyrtus tomentosa*) menggunakan Metode DPPH. *J Scientia*. **8** 37 – 43.

# Tool wear analysis of ceramic cutting tools in the turning of gray cast iron materials

**S Lubis[1*], S Darmawan[1], Rosehan[1], W Winata[1], M Zulkarnain[2]**

[1] Departement of Mechanical Engineering, Faculty of Engineering, Universitas Tarumanagara Jl. Let.Jend S.Parman No.1. Jakarta, 11440 Indonesia

[2] Universiti Kuala Lumpur, Malaysian Institute of Marine Engineering Technology, Dataran Industri Teknologi Kejuruteraan Marin, Bandar Teknologi Maritim, Jalan Pantai Remis, 32200 Lumut, Perak, Malaysia.

*E-mail: sobronl@ft.untar.ac.id

**Abstract.** The development of mechanical and physical properties of metal materials is so fast, that it requires cutting tools that are capable of cutting the metal.Cutting tools must have high temperature resistance, high wear and hardness. Ceramic cutting tools have these properties, so they are suitable for use in cutting hard metals. In the metal machining process, especially machining of cast iron which has high hardness and strength, has a strong reason to use the ceramic cutting tool.This research was conducted to determine wear on ceramic cutting tools when cutting cast iron.The study was carried out experimentally using a Mazak CNC lathe.When the turning process is done, the cutting tool cuts the metal cast iron. Every 10 minutes, the turning process is stopped to observe and measure the wear that occurs on the ceramic cutting tool. Observations and measurements are carried out using a digital microscope.The wear criteria are determined if the tool edge ($V_B$) has been worn at 0.3 mm.The results showed that wear at the cutting speed of 200 m / min at 60 minutes is an adhesion process. BUE (Build Up Edge) wear is a buildup of material layers in the chip area near the cutting tool.

**Keywords:** Turning, ceramic insert, tool wear, gray cast iron, build up edge

## 1. Introduction

The cutting tools are the main thing to consider when choosing the cutting process of metal materials. The high hardness of a material requires cutting tools that have high toughness to do the cutting. Ceramics is one of the cutting materials that is currently developing, ceramics have high hardness, strong resistance to high temperatures.

In the process of metal formation, cutting speed is important to consider to increase high production, and relatively low processing time. The use of high cutting speeds, of course, has an impact on decreasing machining time and increasing the number of products produced. But another effect is the wear of cutting tools that occur faster. The use of cutting tools that have a higher temperature resistance so that they can last longer is very helpful. Cutting tools ceramics are one of the cutting tools that have high temperature resistance, higher hardness compared to carbides. So that this cutting tool can be operated at high cutting speeds.

The tool wear is a general finding in the machining process that significantly affects the size of the product, the quality, the efficiency of the operation of machining, judgments of production and financial feasibility. The wear of the insert have a great influence upon economy of machining. [1].

Cast iron work piece material is a metal material that has high hardness, but also has fragile properties. Cast iron is one type of metal that has been used for a very long time, found by humans among many metals. This metal is widely applied, around 80% for vehicle engines made of cast iron. Cast iron is basically the eutectic iron and carbon alloy. With a melting point around 1200 $^0$C. The low melting point is very profitable, because it is easily liquidated, so that the use of fuel for the metal smelting process is more efficient and low cost. For machinery in austempered ductile iron (ADI), the most widely used type of ceramic cutting tool is the alumina-based class.

There are different classes of ceramic materials for cutting tools, each with different properties. There are two main types of ceramics for cutting tools namely aluminum oxide ($Al_2O_3$) -based, and silicon nitride based ($Si_3N_4$). $Si_3N_4$-based ceramics are formed by $Si_3N_4$ crystals with the intergranular $SiO_2$ phase, sintered with alumina. [2]. Compared to the $Al_2O_3$ class, $Si_3N_4$ based ceramics have higher toughness (except for whiskey $Al_2O_3$ reinforced ceramics), greater hardness, increased thermal shock resistance, and good chemical stability.

Ceramic cutting tool wear behavior must be properly understood for effective use in hard machining of materials. Several studies have been conducted focusing on ceramic cutting tool materials for machining different working materials.[3] Studied the fabrication of materials and the performance of ceramic cutting tools based on $Al_2O_3$ materials /TiC, El-Wardany et al. observed that plastic deformation of ceramic cutting edges, triggered damage to surface roughness, and finally, edge fractures, while machining steel was hardened using ceramic cutting tools.[4]. Chakraborty et al conducted an experiment with cast iron machining using several types of ceramic and carbide cutting tools. Their results show that the dominant wear mechanism that occurs in ceramic-based carbide cutting tools is abrasion, whereas for silicon-based silicon cutting ceramic cutting tools the basic wear mechanism is diffusion.[5].

In metal cutting, the interactions among tool, chip, and workpiece always cause tool wear and other damages to the tool, such as plastic deformation, chipping, and thermal and mechanical cracks [6].
A comparison with tool life and surface roughness of the workpiece produced, with the aim of understanding the wear mechanism that occurs in ceramic cutting tools based on silicon nitride materials and layered carbide cutting tools in the formation of cast iron materials has been pointed by others [7].

Low and high temperatures during the machining process can affect the life of cutting tools [8]. Wear of cutting tools certainly also affect the performance of the tool which has an impact on the value of the resulting surface roughness. [9].

The type of damage that occurs in the cutting tool is plastic deformation caused by high temperature pressure in the active area of the cutting tool so that the hardness and strength of the cutting tool decrease with increasing cutting temperature. An increase in temperature at the cutting tool can occur in the area of the chip or in the main area of the cutting tool. Because of that the shape and location that wear out on the chip area are called crater wear, and wear on the main area of the cutting tool is called edge wear. The type of wear on the cutting tool is shown in Figure 1

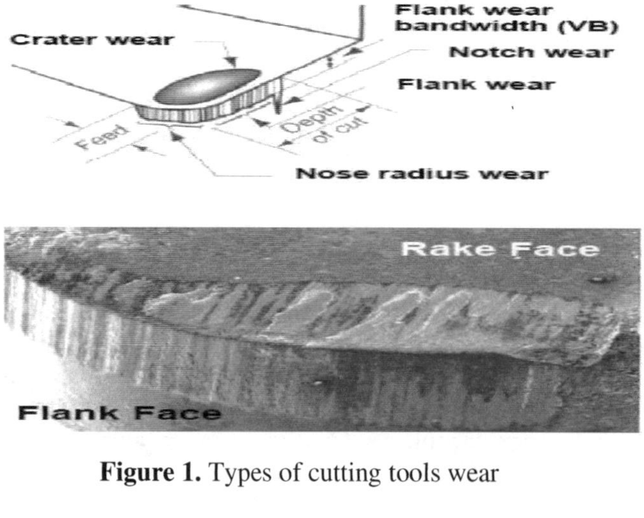


**Figure 1.** Types of cutting tools wear

The flank wear was noticed to be lower than 0.3 mm for both tools. Abrasion and adhesion are most acting wear mechanisms for coated carbide whereas abrasion is most dominant for the ceramic tool. For both tools, speed is the major dominating character for wear at the flank surface [1,10].

## 2. Methodology

The CNC lathe "Mazak Mazatech Quick Turn 8N" was used in this study. The workpiece material that is cut is gray cast iron with a diameter size: 50 mm, length: 150 mm. The tool used is the ceramic cutting tool type TNMG 160404

**Figure 2.** Ceramic cutting tool

Observations and measurements of cutting tool wear were carried out using the Jenco Digital Microscope and surface tester are presented on Fig. 3 and Fig. 4 respectively.

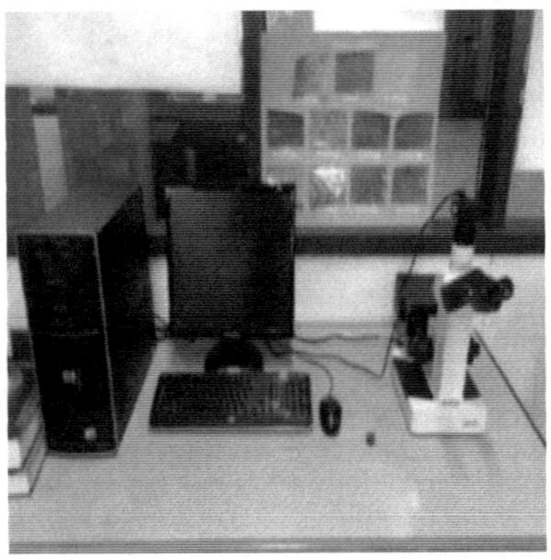

**Figure 3.** Jenco Computer and Digital Microscopes.

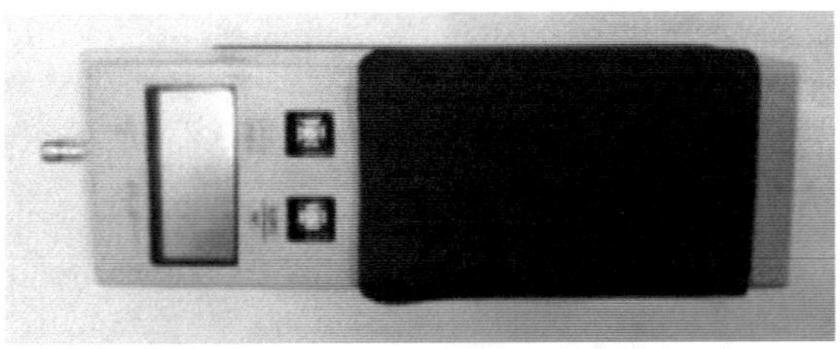

**Figure 4.** Surface Tester

The first step in this experiment is to determine the cutting parameters used. The cutting parameters include cutting, feeding depth and cutting speed. Three cutting speed variations are 160, 200 and 240 m / min with a feeding rate of 0.05 mm / rev, cutting depth of 0.5 mm. The machining process is carried out without using coolant. Then move the cutting tool to the surface of the workpiece. Cutting tool then cuts metal. Machining is stopped for each 10 minute span to see and measure the cutting tool's wear, the edge wear criteria are set at 0.3 mm. If the cutting tool has experienced edge wear, which can be observed and measured using a digital microscope, the cutting tool can no longer be used for metal cutting. However, if the wear value set at 0.3 mm has not been achieved, then the cutting tool can continue the machining process. Besides that, the symptoms of wear / damage of the cutting tool can be known by the vibrations generated on the machine tool or the condition of the work piece surface produced.

## 3. Results and discussion

The observations and experimental measurements of the machining process on the effect of cutting speed on the wear of ceramic cutting tools are presented in the following figure

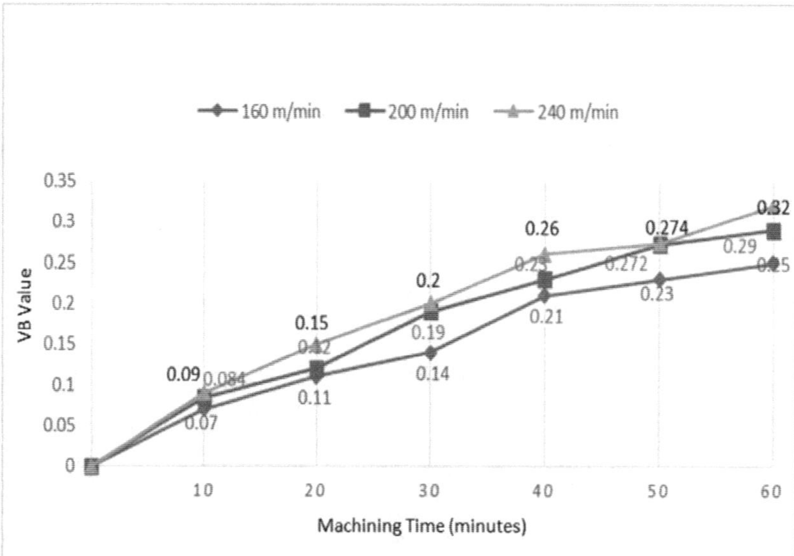


**Figure 5.** Ceramic cutting tool wear

Based on the graph above it can be seen that continuously there is an increase in the value of cutting tool wear so that it reaches the set limit as a limit to the wear of VB cutting tools by 0.3 mm. This happens at the time of machining 60 minutes. Wear occurs of course because of the friction in the contact area of the cutting tool and the workpiece causing thermal on the surface of the cutting tool, the longer, the heat that occurs in the cutting tool contact area has increased. This certainly affects the crystal structure of the cutting tool material until it finally undergoes abrasive and wear on the edge of the cutting tool. This happens as an effect of increasing cutting speed (Vc). Can be seen from the graph at the cutting speed of 160 m / min at 60 minutes the wear growth that occurs is very slow compared to the cutting speed of 200 m / min and 240 m / min to 60 minutes, this is because the cutting speed value used is low so the heat that occurs is not affect the overall contact area of the cutting tool, however fine erosion has also occurred due to the cutting force that occurs when the cutting tool cuts the metal workpiece.

Topography of ceramic cutting tool edge wear at 60th minute machining with cutting speed variation (Vc) is presented on Fig. 6 to Fig. 8 respectively.

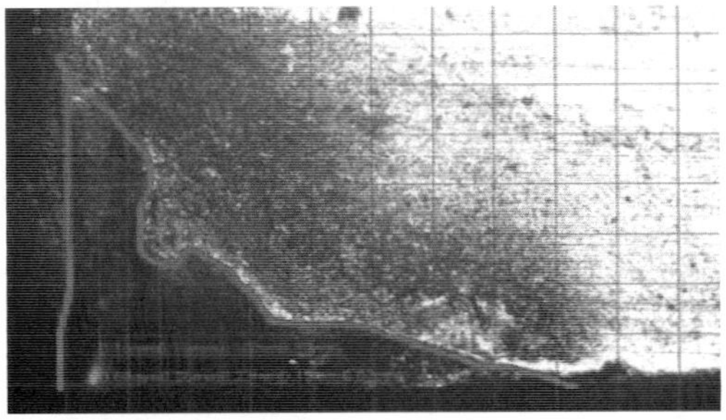

**Figure 6.** Flank wear at cutting speed 160 m / min

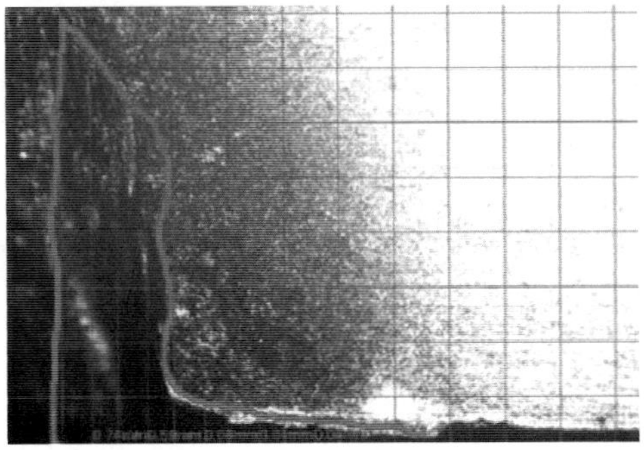

**Figure 7.** Flank wear at cutting speed 200 m/min

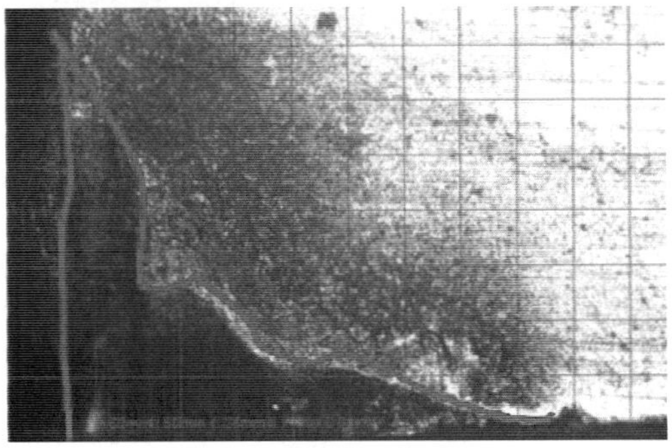

**Figure 8.** Flank wear at cutting speed 240 m/min

During the machining process is carried out by using a cutting speed of 160 m / min minutes, observations on the wear of the cutting tool are carried out, and the wear value is obtained (VB) in the 60th minute. main cutting tool. The abrasive wear continues to rise in the main areas of the cutting tool. Whereas in the main field of the cutting tool, this abrasive process causes the formation of edge wear, as a result the work surface surface roughness increases, and in the machining process, there is increasing noise. Wear due to this abrasive process continues to grow to reach the cutting edge of the cutting tool. When the cutting speed is increased by 200 m / min, the wear of the cutting tool starts to increase in the 60th minute. This is due to high friction due to increased cutting speed causing higher heat, so that the heat distribution at the cutting tool contact angle occurs faster which causes abrasive and in the end part of the chisel is eroded, wear and tear occurs. Is BUE (build up edge) which is a buildup of material layers in the chip area near the cutting tool.

The accumulation of layers of material that has just formed is attached to around the main field. This wear mechanism is due to the relatively high pressure and speed that causes the newly formed workpiece surface to stick with the cutting tool's angle to come off with the chip.

The use of cutting speed 240 m / min, wear that occurs in the 60th minute. The type of wear that occurs is plastic deformation. This mechanism occurs because the cutting tool is deformed plastic due to compressive load and deformation due to the high shear load in the main cutting tool causing edge wear. This is due to increasing cutting times and high cutting speeds so that the power of the cutting tool decreases.

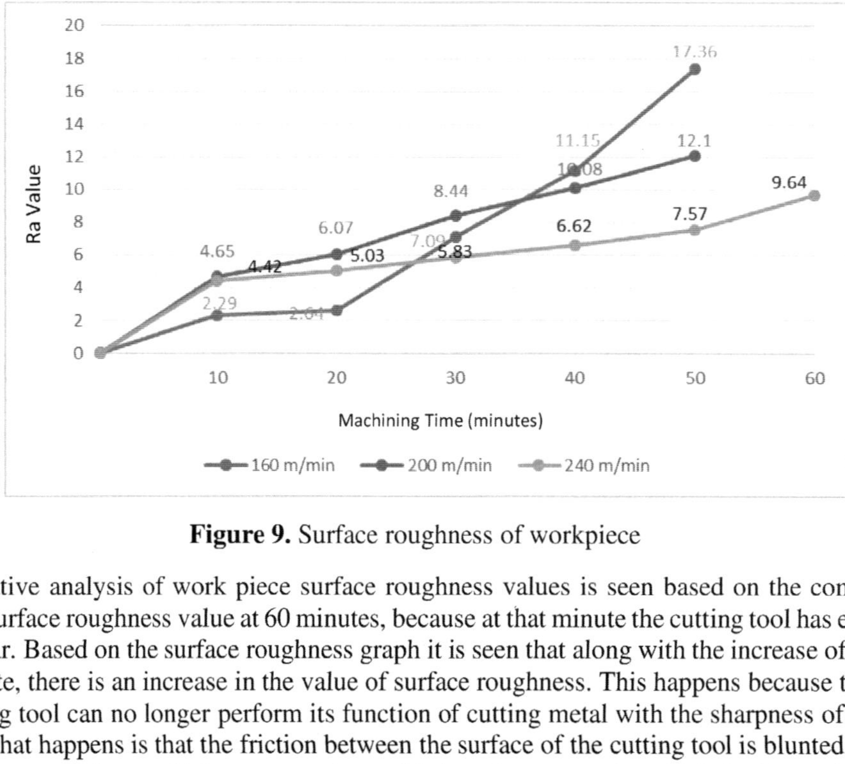


**Figure 9.** Surface roughness of workpiece

Comparative analysis of work piece surface roughness values is seen based on the comparison of workpiece surface roughness value at 60 minutes, because at that minute the cutting tool has experienced wear and tear. Based on the surface roughness graph it is seen that along with the increase of the cutting tool wear rate, there is an increase in the value of surface roughness. This happens because the wear tip of the cutting tool can no longer perform its function of cutting metal with the sharpness of the cutting angle, but what happens is that the friction between the surface of the cutting tool is blunted against the surface of the workpiece, splitting occurs due to friction that causes heat high in the cutting tool and workpiece contact area, due to a decrease in material quality, the debris is formed due to the push of the cutting tool [10-11].

From the observations it can be seen that the greater wear of the cutting tool is directly proportional to the increase in the value of the surface roughness of the workpiece. This is because when cutting, the cutting tool's corners are no longer cutting, but rather because of cutting caused by friction from the worn cutting tool. This shows that the effect of the wear of the cutting tool causes damage to the surface of the workpiece.

## 4. Conclusion

From the results of testing and analysis carried out, it can be concluded that Increasing the cutting speed has an effect on wear on the edge of the cutting tool, so the cutting tool becomes shorter. The roughness of the workpiece (Ra) is directly proportional to the value of the cutting tool (VB) wear, because in the cutting process, the cutting tool's cutting edge is no longer cutting, but because of cutting caused by friction from the worn cutting tool. The more worn the cutting tool and the lower cutting speed, the higher the work piece surface roughness produced. Wear that occurs in the cutting tool is an abrasive. The wear of the cutting tool in the cast iron machining process takes place in the 60th minute

## References

[1] Panda A, Sahoo A K, Rout A K, Kumar R and Das R K 2018 Investigation of Flank Wear in Hard Turning of AISI 52100 Grade Steel Using Multilayer Coated Carbide and Mixed Ceramic Inserts *J. Procedia Manufacturing.* **20** 365-371.

[2] Yeckley R 2005 *Ceramic Grade Design* in: Kennametal Comprehensive Application Engineering Guide Kennametal University Latrobe PA 12 2–12

[3] Xu CH, Ai X and Huang C Z 2001 Fabrication and Performance of an Advanced Ceramic Tool Material *J. Wear.* **249** 503-508

[4] El-Wardany T I, Mohamed E, Elbestawi M A, 1993 *Material Side Flow in Finish Turning of Hardened Steel with Ceramic Tools, Contact Problems and Surface Interactions in Manufacturing and Tribological systems* New York: ASME PED 67 159–170.

[5] Chakraborty J A, Bhaduri S B and Mahajan Y R 1990 Performance of Ceramic Cutting Tools in Machining Cast Iron. *Powder Metal Int* 22

[6] Lubis S, Rosehan, Darmawan S and Adianto Malik R 2019 The Influence of Cutting Speed Variation in Turning of AISI 304 Materials on Wear and Tool Life Coated Carbide Cutting Tools. *Int. J. of Mechanical Engineering and Technology (IJMET)* **10** 203-210

[7] Diniz A E and Ferrer J A G 2008 A Comparison between Silicon Nitride-Based Ceramic and Coated Carbide Tools in the Face Milling of Irregular Surfaces. *J. of Materials Processing Technology.* **206** 294–304.

[8] Kalpakjian S 1991 *Manufacturing Processes for Engineering Materials* 2nd ed (Chicago: Addison-Wesley Publishing Company Inc)

[9] Sobron M L, Darmawan S and Adianto Ericsen 2019 Effect of Cutting Speed on Temperatutre Cutting Tools and Surface Roughness of AISI 4340 Steel IOP Conf. Serises: *Materials Science and Engineering* **508** 012053

[10] ASM Handbook 1989 *Metallography and Microstructures* 16 Machining 27–31.

[11] Mrihrenaningtyas and Prayadi R 2015 Analisis Umur Pahat dengan Variasi Sudut Geram, Kecepatan dengan dan Tanpa Pendingin. *Pros. Seminar Nasional Sains dan Teknologi Terapan* III (Teknologi Adhi Tama Surabaya)

[12] Rochim T 1985 *Teori dan Teknologi Proses Pemesinan* (Bandung: Jurusan Teknik Mesin, Fakultas Teknik Industri - Institut Teknologi Bandung)

# The effect of process water recycle on hydrothermal treatment of yard long bean (*Vigna unguiculata ssp. sesquipedalis*) and water spinach (*Ipomoea reptans*) seeds

**D Kurniasari[1], A T Yuliansyah[2] and C W Purnomo[2*]**

[1] Magister of System Engineering, Faculty of Engineering, Gadjah Mada University
[2] Department of Chemical Engineering, Engineering Faculty, Gadjah Mada University

[*] Email: chandra.purnomo@ugm.ac.id

**Abstract**. Hydrothermal treatment is one of thermal biomass conversion method that the solid product can be used as fuel. The main problem of hydrothermal system is the liquid product that usually become waste. The water requirements of hydrothermal treatment need to be assessed to determine commercial feasibility. This research aims to study the influence of reusing liquid product to become process waster in hydrothermal treatment of yard long bean and water spinach seed with temperature variation of 180, 200 and 220 ℃, biomass and water ratio of 1:5, and residence time of 40 minutes. Solid product was characterized in terms of mass yield and heating value (HHV). It is found that the solid yield value increased 8% from the first to the sixth process using recycle water. The biggest HHV of solid product was the hydrothermal process with temperature of 220℃ using the 5$^{th}$ recycle liquid at 5499 cal/g for water spinach seed and 5466 cal/g for long bean seed.

**Keyword:** biomass seeds, water recycle, hydrothermal treatment.

## 1. Introduction:

Biomass is an abundant renewable energy source in nature and it is considered one of the most promising alternatives to fossil fuels for the production of bio-fuels. One of the large available biomass in Indonesia has been not widely studied is vegetable seeds such as yard long bean (*Vigna unguiculata ssp. sesquipedalis*) and water spinach (*Ipomoea reptans*) seeds. The statistical data of Indonesian Agriculture show that the amount of yard long bean production was 313 tons, and 3980 ton for water spinach seed in 2017 [1]. The seed industry will sell the seeds that pass the selection section. In the selection process, at about 10% seed will not pass the quality control and become solid waste. Since seed is hard and have strong mechanical integrity it is difficult to be decomposed and recycle. By hydrothermal carbonization (HTC), biomass could be converted into a solid fuel, liquid and condensat [2].

Usually, hydrothermal treatment processing uses temperature from 180 to 320℃ in suspension with pressurized water for a period time [3] .The aqueous products contain various sugars, organic acids, and furfurals and can be potential feed for valuable chemical recovery. Another studies report that process water from hydrothermal treatment biomass contained fenol, oligomer glucose, alcohol and organic acid [4].The process water of hydrothermal treatment is an essential, a large demand for process water can be minimized by recycle. Recycling process water can reduce wastewater treatment costs, since it has to be treated before discharges into the environment [5]. Previous reports were studied the effect of

IOP Conf. Series: Materials Science and Engineering **857** (2020) 012004 doi:10.1088/1757-899X/857/1/012004

recycle process water of hydrothermal of poplar wood [6], municipal green waste (MGW), and orange pomace [7]. However, these studies still use common solid waste and did not consider the effect of recycling water to the alteration of solid product properties such as heating values.

In this study, the effect of process water recycle from hydrothermal treatment was studied using water spinach seed and ipomoea sp seed with temperature variation of 180, 200, 220°C with biomass and water ratio of 1:5, residence time of 40 minutes and procedure was repeated ten times at 180°C, 200°C and nine times at 220°C. The hydrochar was analyzed in terms of mass yield and heating value (HHV).

## 2. Material and methods

### *2.1 Material*

The long bean (*Vigna unguiculata ssp. sesquipedalis*) and water spinach (*Ipomoea reptans*) seeds were collected from a seed industry (East West Seed Indonesia) in Purwakarta, West Java, Indonesia. The seed were collected from off spec. portion (rejected seed) and characterized for lignocellulosic content and shown in Table 1.

**Table 1. Lignocellulosic characteristic of the rejected seeds**

| Seed variety | Cellulose, %wt. | Hemicellulose, %wt. | Lignin, %wt. |
|---|---|---|---|
| Yard long bean (RKP) | 18.1 | 35.1 | 7.4 |
| Water spinach (RBK) | 29.5 | 30.5 | 10.2 |

### *2.2 Experimental procedure*

Hydrothermal treatment process were conducted in 2L batch type autoclave reactor with a shaking system, temperature controller and pressure control. For each of the intial runs, a mixture of seed and demineralized water was loaded into the glass liner with a mass ratio of 1:5 of biomass:water (100 gr biomass : 500 mL water) and maximum temperature was set on 180, 200 or 220°C. The residence time during processing temperature was fourty minutes and then reactor was naturally cooled and after that the solid product was collected by filtration. The hydrothermal reactor scheme is provided in Figure 1.

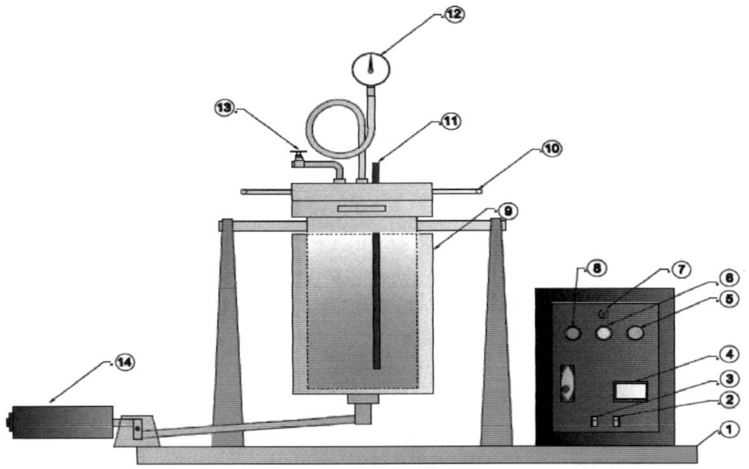


Note:
1. Chasis
2. On or Off switch
3. Temperature controlller
4. Heater Switch
5. Heater Indicator
6. Motor Indicator
7. Alarm
8. Power Indicator
9. Reactor
10. Reactor switch
11. Stirrer
12. Pressure gauge
13. Pressure valve
14. Motor shaker

**Figure 1.** Hydrothermal reactor scheme

The solid products was dried in oven with temperature of 180°C for three hours. The hydrochar was analyzed to obtain heating value and the liquid products was measured for its pH using Hanna instrument. Some liquid by the volume of 250 mL was separated and mixed with deionized water to become 500 mL of liquid mixture and then used for the next hydrothermal batch. The solid yield can be calculated using the following equation

$$Y\ (\%)\ = \frac{Wdb\ Product}{Wdb\ feed} \times 100\% \qquad (1)$$

where, Y is solid yield (%) and Wdb is mass of dried solid basis in grams.

For ease of identification the sample code for long bean is RKPx and for water spinach is RBKx with x is the recycle time of the process water, while for the first process using all fresh water is x=0. For example, the data of RKP3 means the hydrothermal process of yard long bean seed using the third recycle of process water.

## 3. Results and Discussion

Hydrothermal treatment is a process of thermal degradation of solid to form gases, liquids and solid products. In addition, the thermal degradation can decompose long chain molecules into simple structures with some of these molecules are dissolved into liquid and dispersed into gases and residues [8]. Commonly, the water demand for hydrothermal process is always use clean water in every batch. This can cause a huge quantity of wastewater since the water requirement in every batch is at about 5 times larger than the solid. In this study, hydrothermal process water is recycled at around 10 times.

The first batch of liquid product is collected and mixed with fresh water (1:1) and reused for the second batch and so on. In every batch the seed is always fresh only the water is half recycled. The solid yield of the seeds hydrothermally treated with water recycle is shown in Figure 1 and 2. In general, the yield of solid for RBK is higher than RKP this because the higher content of cellulose and lignin in water spinach seed that will contribute to the char production.

The solid products from long bean seeds increases from the first to the sixth recycle. Meanwhile for the other seed, the increase in solid products occured until the third recycle. After the increase, the solid yields are mostly decrease by the recycling progression. The yield of cowpea seeds with a temperature of 180°C increased from the first recycle process by 53.42% to 62.08% in the sixth recycle and then remain steady to the end of recycle. This increase of solid product is caused by compounds accumulation in the hydrothermal treatment process water from the previous processes. This results was confirmed by Uddin et al. [9] which shows that the optimization of the recycle of process water was very important because biomass pretreatment could be influenced by the presence of organic and inorganic materials remaining in the liquid product. An increase in the mass of solids could be generated from the deposition of compounds such as dissolved sugar from the decomposition of hemicellulose and cellulose that came out during the initial hydrothermal process.

After experiencing an increase, most of the solid yields are decrease in the later recycle. The hydrochar solids of water spinach with treatment temperature of 200°C has a peak at 61.56% in sixth recycle and then decrease to 55.46% in the last run. Meanwhile, the hydrochar yield of long bean in the same temperature has an optimum level in the third recycle at 55.48% and decrease to 47.76% in the last process. The decrease in solid yield could be due to the increase of pH in the hydrothermal solution by the progressing of the recycle sequence. The increase of pH will accelerate the solid decomposition and catalyze the compound cracking. Thus, somehow there is a kind of optimum recycling step for obtaining the solid yield.

In Heidari et al. [10] study, by recycling hydrothermal process water it had a positive effect on hydrochar yield value, the yield increased after the first recycle was 10.26%, while, the second recycle the increase in yield value was only 1.65% from the first recycle. On the other hand, Tercero et al. [11] stated that the amount of recycled water process carried out has not significantly affected the increase in hydrochar yield, but the yield of hydrochar value was significantly affected by temperature. Other reported that the decrease in hydrochar was caused by the competition reaction of decomposition and depolymerisation of cellulose and hemicellulose content [12]. In addition, hemicellulose compounds were almost completely hydrolyzed at a temperature of 180°C while lignin was hydrolyzed at 200°C and cellulose was not hydrolyzed at temperatures below 220°C. Lignin is a lignocellulose compound which is very difficult to be degraded since it has a stable phenolic structure that allows it to resist decomposition into liquids and gas fractions [13].

According to Kumari[14] hydrolysis occured because of the formation of hydronium acid ions ($H_3O^+$) and alkaline hydroxide ions ($OH^-$) derived from ionization of water as reactants. The higher the reaction temperature used, the higher the ion dissolved in the water.

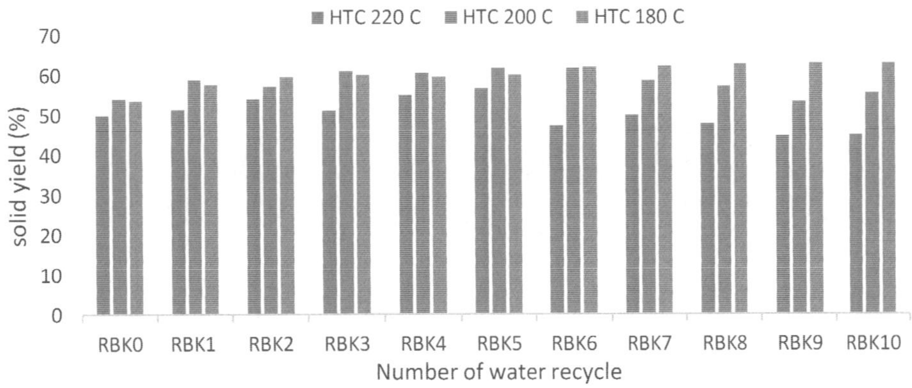


**Figure 2**. Yield of hidrochar products from water spinach seeds

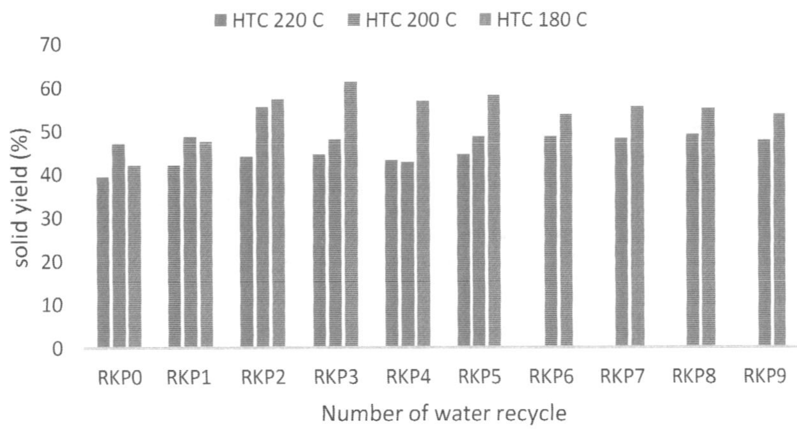


**Figure 3**. Yield of hydrochar products from yard long bean seeds

In this study, HHV was analyzed for char at early, middle and end of recycle sequence as shown in Figure 4 and 5. In general, the solid product of water spinach seed has a higher heating value by the increase of temperature, with the highest value of 5499.0 cal/g at the 5$^{th}$ recycle, but slightly reduced at the last recycle. Whereas the heating value at 200°C increased from 4807.7 cal/g on the first recycle to 5284.7 cal/g fifth recycle and slightly increasing at the end step. The similar trend also happened to the HHV of long bean chars with the highest HHV is for HTC 220 at 5465.6 cal/g.

The increasing of HHV at the middle of recycle number has a similar correlation with the solid yield. Thus, it can be said that the optimum solid product as a solid fuel with high yield and calorific value can be reach by the fifth recycle of process water.

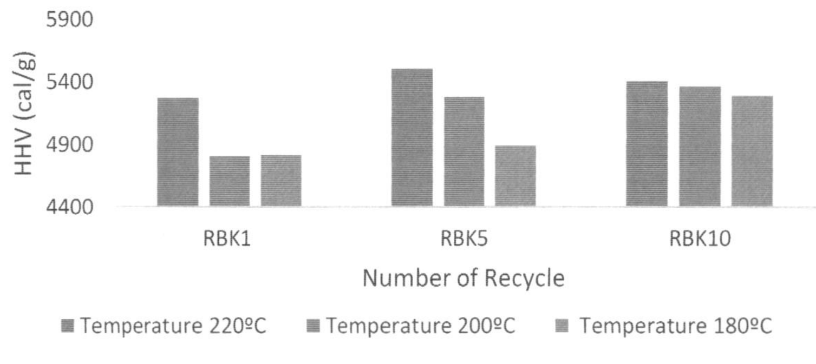


**Figure 4.** Higher heating value of hydrochar from water spinach seeds

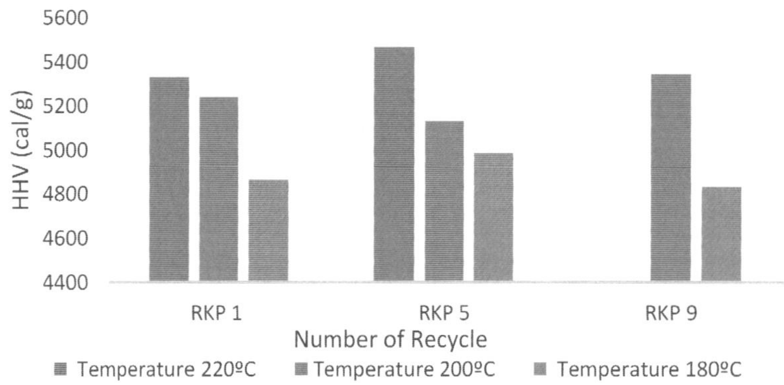


**Figure 5.** Higher heating value of hydrochar from yard long bean seeds

The energy content of hydrochar was influenced by water, ash, volatile and fixed carbon content. Thermal decomposition process or devotalization which was influenced by increase of the HTC temperature will remove the volatile content and leave the ash and carbon causing creating solid with high calorific value [15].

Uddin et al. [9] stated that the recycled water gives an influence on increasing the heating value of about 4% increased from the initial process which could be due to the content of organic acids such as acetic acid produced from previous decomposition. More acidic solution will fasten the decomposition process and removal of volatile content thus increase the heating value. The liquid products pH for selected sample is provided in Table 2.

**Table 2. pH of some liquid product with number of recycle process at 220°C**

| Seeds | Liquid pH of number of recycle | | | |
|---|---|---|---|---|
| | 1 | 5 | 7 | 9 |
| RBK | 4.2 | 4.1 | 4.0 | 4.2 |
| RKP | 4.6 | 4.2 | 4.3 | 4.5 |

Heidari et al. [10] stated that the amount of acid components formed in process water was due to decomposition and depolymerization reactions of biomass when the temperature rises. The solid mass increased when pH was low, based on the results obtained the highest acidity level occured in bio-oil from both seeds. Reza et al. [16] stated that liquid will increase the acidity with the same time as the sugar content also increases in liquid products. Sugar contents can increase the solid yield through condensation and carbonization [17]. The progression of water recycle lead to the saturation with the balance of acid compounds, as demonstrated by Biller et al. [18] where the pH had increased from 7.6 in the initial recycle of process water to pH 10 at the final recycle.

## 4. Conclusion

Process water recycles of hydrothermal treatment could give effect to increasing solid mass yield as well as heating value of the solid product. This effect is mainly due to the content inside the recycled liquid by the combination of acidic and sugar compounds. The optimum recycle has to be determined for each waste in order to get the highest solid yield and calorific value of the solid product.

## References

[1] P. D. and S. I. Pertanian, *Statistik Pertanian 2017 Agricultural Statistics*. 2017.
[2] M. Titirici, C. Fühner, O. Bens, J. Kern, and K. Emmerich, 2011, **2**, pp. 89–124.
[3] S. Román, J. M. Valente Nabais, B. Ledesma, J. F. González, C. Laginhas, and M. M. Titirici, , 2013, *Microporous Mesoporous Mater.*, **165**, no. Complete, pp. 127–133.
[4] C. Li, G. Zhang, Z. Zhang, D. Ma, L. Wang, and G. Xu, 2015 *Chem. Eng. J.*, **279**, pp. 530-537.
[5] J. Stemann and F. Ziegler, '19th European Biomass Conference and Exhibition, 6-10 June 2011, Berlin, Germany', no. June, pp. 6–10, 2011.
[6] J. Köchermann, K. Görsch, B. Wirth, J. Mühlenberg, and M. Klemm 2018 *J. Environ. Chem. Eng.*, **6**, no. 4, pp. 5481–87.
[7] E. Erdogan *et al* 2015 *Bioresour. Technol.*, **196**, pp. 35–42.
[8] A. T. Yuliansyah, T. Hirajima, S. Kumagai, and K. Sasaki, 2010. *Waste and Biomass Valorization*, **1**, no. 4, pp. 395–405.
[9] M. H. Uddin 2016 *Environ. Prog. Sustain. Energy*, vol. 00, no. 00, pp. 1–10.
[10] M. Heidari, S. Salaudeen, A. Dutta, and B. Acharya 2018 *Energy and Fuels*, **32**, no. 11, pp. 11576–86.
[11] E. A. Ramos-Tercero, A. Bertucco, and D. W. F. Brilman 2015 *Energy and Fuels*, **29** no. 4, pp. 2422–30.
[12] H. Singh, K. Jamie, and M. Animesh 2017 *Waste and Biomass Valorization*
[13] A. Demirbaş 2001 *Energy Convers. Manag.*, **42**, no. 2, pp. 183–188.
[14] R. Kumari and K. Pramanik 2013 *Appl. Biochem. Biotechnol.*, **171** pp. 771–785.
[15] A. Abidin, C. W. Purnomo, and R. B. Cahyono 2018 in *AIP Conference Proceedings 2018*, **2026**, no. 1, p. 020049.
[16] M. T. Reza *et al.* 2014 *Appl. Bioenergy* **1**
[17] A. B. Ross, P. Biller, M. L. Kubacki, H. Li, A. Lea-Langton, and J. M. Jones 2010 *Fuel*, **89**, pp. 2234–43.
[18] P. Biller, R. B. Madsen, M. Klemmer, J. Becker, B. B. Iversen, and M. Glasius 2016 *Bioresour. Technol.* **220**, pp. 190–199.

# Development and characterization of porous hydroxyapatite-Alumina composite for engineering application

**A Arifin[1], Gunawan[1]*, A Priyadi[1], F A Sanjaya[1]**

[1] Department of Mechanical Engineering, Faculty of Engineering, Universitas Sriwijaya, Indralaya, 30662, Sumatera Selatan, Indonesia

* Email: gunawan@unsri.ac.id

**Abstract**. Synthetic hydroxyapatite is a material that has a chemical composition and molecular structure similar to bones and teeth, generally this material is widely used for medical applications including as an implant material. Non-medical applications of porous HA include as catalyst in chemical reaction, packing media for column chromatography, and as probe for gas sensors. The study aims to develop porous material based HA/$Al_2O_3$ composites. HA/$Al_2O_3$ composites prepared with three variations of latex space holder weight fractions namely 10%, 20%, and 30% and using the unidirectional compaction method to produce composites. Two step holding process sintering process applied on the green body specimens. Firstly, hold for 60 minutes at temperature of 600°C and secondly 3 hours at final temperature of 1200°C with 5°C/min heating rate. The apparent density test was performed based on the Archimedes' principle. Porosity characteristics of the sample show that density of the composite decreased with increased space holder. Compressive strength results show that the addition of SH weight fraction decreased the compressive strength of porous hydroxyapatite-$Al_2O_3$ composites. The morphology observation via SEM showed pore formation due to burning of space holder and interconnecting porous formed.

**Keywords:** porous, hydroxyapatite, alumina; space holder, powder metallurgy

## 1. Introduction

Hydroxyapatite (HA) synthetic with calcium and phosphor as main composition is having excellent biocompatibility and osteoconductivity properties. Both natural properties of synthetic HA material is main reason of using it as bone grafting material in hard tissue implants in medical applications [1]. Generally porous HA ceramics for non-medical applications used as catalyst in chemical reaction, packing media for column chromatography, and as probe for gas sensors [2]. For catalyst application, in hot water environment about 80°C, HA increased ribose formation [3,4]. It is believed that the effect of spacing of calcium ions on the external surface of HA effectively enhanced chemical reactions.

HA with chemical composition $Ca_{10}(PO_4)_6(OH)_2$ can be produced from synthetic materials and natural resources. HA prepared from natural sources including animal bones, corals, eggshells, etc. [4]. The advantages of syntheses HA from natural sources are simpler and cheaper since no chemicals addition. Several methods to produce porous hydroxyapatite established in literatures [5][6][7]. The aim of many of these methods is the manufacture of porous HA, with controlled percentage of porosity without decreasing its mechanical properties. One of the method is by applying a material as space holder in powder metallurgy methods to produce porous material that offers flexibility and it is also a cost-effective alternative.

In this case, in the conventional method several parameters must be considered to limitation in controlling the quantity, size, distribution and pore morphology. In addition, in an effort to obtain efficient process in terms of cost factors, reproducibility and flexibility compared to new technologies such as field assisted sintering technology (FAST), ion beam milling, laser sintering, etc. must be considered. In contrast, applying a material as space holder in powder metallurgy methods techniques provide suitable routes for obtaining porous HA structures; these method also have extraordinary advantages including cost-effective and non-toxic methods, without toxic agents that be able to affect porous purpose

Space holder is short-term constituent part mixed to the matrix. The porosity formed from constituent burnt in the sintering process and leave marks. Some types of space holder material have been used including polymethyl methacrylate, NaCl, $NH_4HCO_3$, corn and tapioca starch [8].

Hydroxyapatite intrinsically owing to low mechanical properties such as compressive strength and bending strength, therefore it is necessary to improve its mechanical properties by reinforcing with $Al_2O_3$. The main objective of this work is to characterize fabricated porous hydroxyapatite-$Al_2O_3$ composites with the application of space holder materials (i.e. latex powder) through the powder compaction method.

## 2. Materials and Method

In the research process Hydroxyapatite (HA) material is derived from cattle bones obtained from waste or residual bovine bones in Palembang. The bones collected are femur bones. The raw material for reinforcement in the form of $Al_2O_3$ (Alumina) is obtained from the Online Store and the raw material for the space holder in the form of latex powder is obtained by crushing the rubber eraser into powder. HA prepared by boiling bovine bones to eliminate of fat contained in bones and to clean the remaining meat and muscles for 4 hours.

Bovine then dried to remove the remaining moisture content and reduced the dimensions by using a cutting grinder. r use as hydroxyapatite must be calcined using an electric furnace at a temperature of 600°C for 2 hours and cooled slowly to match room temperature. Furthermore, the calcined bovine bone is mashed by using mortar by pulverizing it. The result of mashed beef bone powder is then sifted to get the desired particle size. Afterward calcination process on electric furnace at 600°C and 3°C/min heating rate for 30 minutes before cooled inside furnace. During calcination process the organic substances decomposed and water evaporated. Furthermore, the calcined bovine bone is mashed by using mortar by pulverizing it.

The result of mashed bone powder is then sieved to get the desired particle size. The production of porous HA/$Al_2O_3$ composites was carried out by mixing hydroxyapatite powder (200 μm), alumina powder (55 μm) and space holder powder with composition 75%:25% of weight fraction. Subsequently space holder of latex powder (200 μm) added to composite with 10%, 20% and 30% weight fraction variations of total composite weight. The weighed powder was then mixed for 1 hour using a ball mill with a spin speed of 225 rpm using a Groschopp Viersen FRG ball mill.

The mixed powder then is loaded into a die (die) cylindrical samples with length 12 mm and diameter 10 mm for the compacting process with a pressure of 2000 Psi for 10 minutes. After the compacting process is carried out, then the specimen is removed from the mold. The sintering process is carried out in conventional electrical furnace Nabertherm GmbH. For adequate sintering process complete burn of latex powder facilitated via two step holding process (fig. 1). First, hold for 60 minutes at temperature of 600°C and secondly 3 hours at final temperature of 1200°C with 5°C/min heating rate.

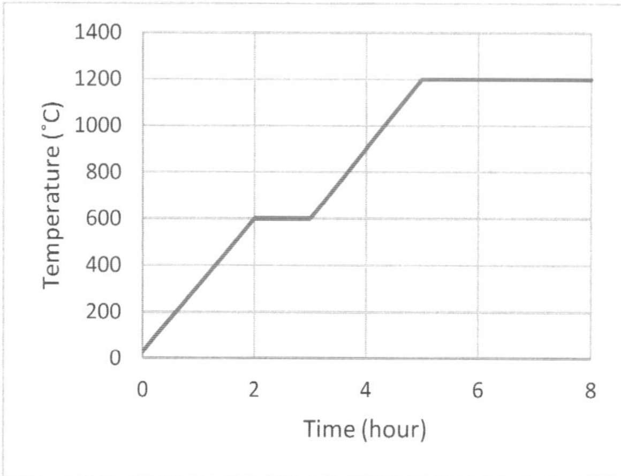


**Figure 1.** Heating schema in sintering process

Various experimental method has been done to characterize powder green bean, HA and $Al_2O_3$ as well as porous HA- $Al_2O_3$ composite including SEM evaluation using FEI Inspect S50, analyze decreasing weight of space holder via thermo gravimetric analyzer from TA Instruments TGA Q500, phase formed during syntheses using an X-Ray Diffraction of Rigaku MiniFlex 600 and compression strength tested using Bongshin Hydraulic Universal Material Tester

## 3. Results and Discussion

Thermo gravimetric analyzer (TGA) testing was carried out to determine the rate of weight change in the temperature function of rubber powder from room temperature to a temperature of 600°C as can be seen based on figure 2. Up to the temperature of 140°C the rubber powder material begins to experience a weight reduction caused by evaporation of the moisture content. The decomposition process occurs until the temperature of 300°C that temperature susceptible to a drastic decrease weight from 98.51% to 43.08%. Mass reduction continued to temperature of 580°C rubber powder remained 23.032% by weight where the rubber powder sample has not burned out completely and formed ash.

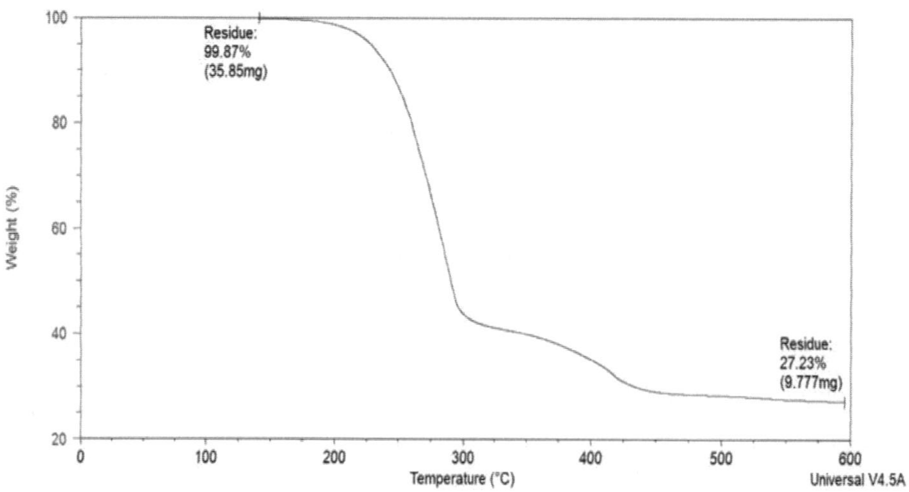


**Figure 2.** TGA result of latex powder

As can be seen from figure 3 (a), the XRD results of calcined bovine bone at a temperature of 600°C in the red graph while hydroxyapatite standard according to ICDD 09-432 in the blue graph. It can be seen that the XRD graph of the sample close to the ICDD standard 09-432 at 2θ of 26.270, 29.61, 32.43, 33.31, 35.41, 40.19, 51.52, and 64.65°. Figure 3.b shows XRD pattern of alumina powder, as it can be seen that the result of XRD match perfectly with peak of alumina according to ICDD 46-1212 card. also XRD result depict high crystallinity of powder alumina.

XRD results on porous HA-Al2O3 sintered at 1200°C are shown in Figure 4 producing a more crystalline hydroxyapatite phase characterized by higher peak intensities than calcined powders. In addition, the β-TCP phase (ICDD 09-169) was also detected due to partial dehydration which caused the decomposition of hydroxyapatite to TCP due to the CaO reaction and water vapor which changed the location of the phase boundary. Third phase detected is the phase of alumina powder however another phase resulted from latex powder is not detected.

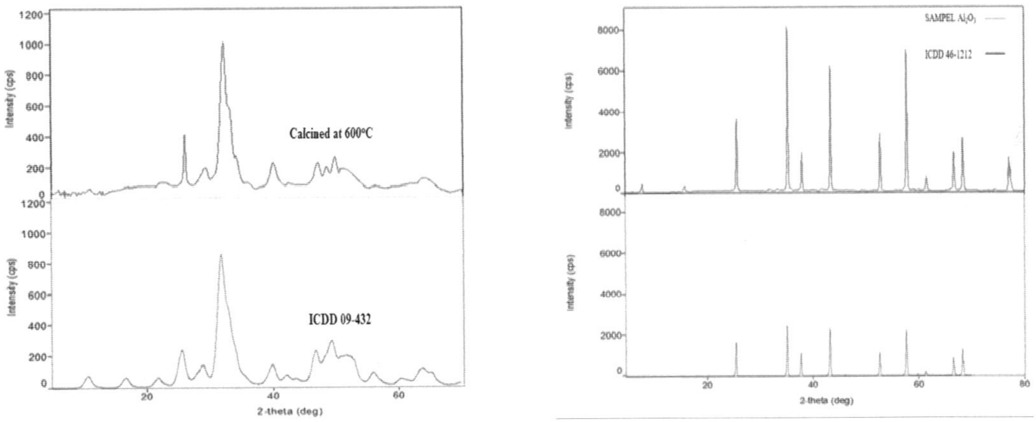


a) b)

**Figure 3.** a). XRD results of calcined bovine bone temperature 600°C
b). XRD result of as received $Al_2O_3$

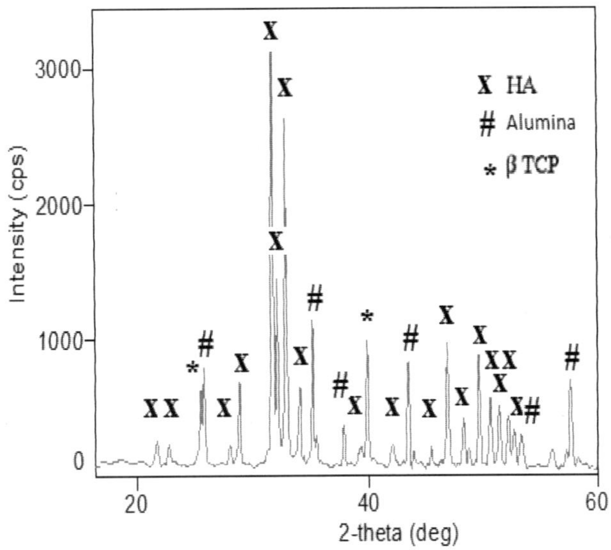


**Figure 4.** XRD result of Porous HA-$Al_2O_3$ Composite Sintering Temperature 1200°C

Figure 5.a depict the average percentage porosity HA/$Al_2O_3$ composite. The higher porosity was obtained in samples with a weight fraction of 10% space holder that was equal to 54.49%. However, the addition of space holder weight fraction up to 30% will reduce its porosity by 45.21%. The trend can be

IOP Conf. Series: Materials Science and Engineering **857** (2020) 012005 doi:10.1088/1757-899X/857/1/012005

confirmed by the TGA results in which the weight percentage of latex powder that remains when burned at 600°C is 27.23% in the form of ash. Increasing the percentage of space holder will accumulate and increase the amount of ash remaining.

The remnants of this ash will fill the space between the grains formed by the burning of latex powder. Increasing percentages the weight of the space holder from 10 to 30 has an undesirable effect on the porosity in term of height porosity material design. The addition of space holders is not able to increase porosity but decreases.

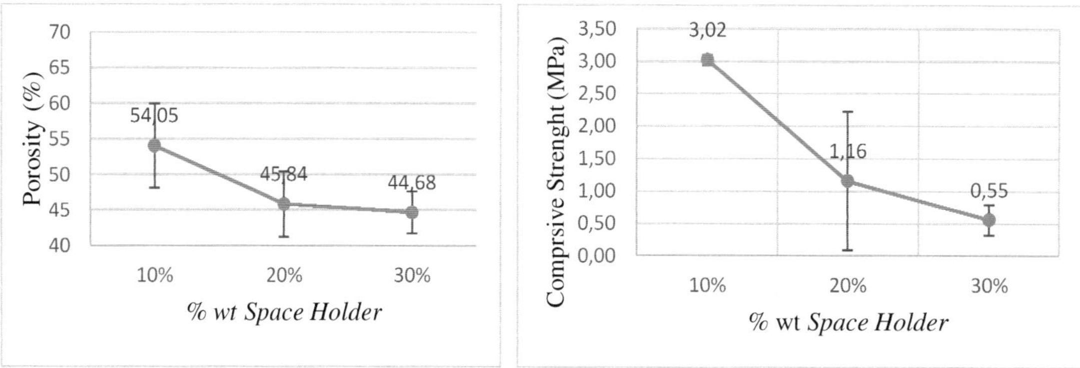


a) b)

**Figure 5.** a) Average porosity correspond to weight fraction of space holder
b) Compressive Strength Results of Composite at Various Weight Fraction of Space holder

The results of compressive strength test of HA-Al2O3 composites can be seen in Figure 5.b. From the results show that the compressive strength at 30% weight fraction space holder is the lowest compared to the variation of 10% and 20% space holder weight fraction. In contrast, the 30% weight fraction has the lowest porosity compared to other samples. The low compressive strength of this 30% sample indicates that accumulation of ash residual latex powder is more numerous than in samples with less space holders. In the ash sintering process the remaining space holder fills the space between HA grain boundary and interferes with the diffusion process between HA particles thereby reducing the amount of diffuse HA particles which in turn reduces the compressive strength.

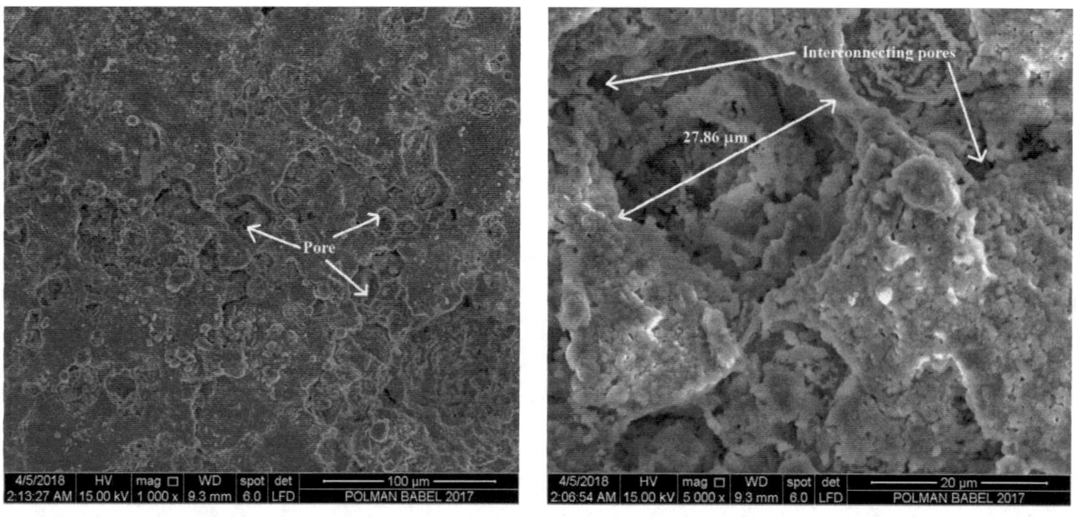


**Figure 6.** SEM Results of HA-$Al_2O_3$ Composite at Two Different Point

Figure 6 shows the observation of microstructure using SEM on a sample of 30% space holder. As can be seen, in two different observation points with magnification of 1000 and 5000, porosity formed has varying sizes with a maximum size of 27.86 µm. The large porous size is formed due to the process of burning down the space holder, while there are also smaller pores associated with particle diffusion in the sintering process. In addition, interconnected porous can also be seen clearly in the sample.

## 4. Conclusion

The conduct of the study as described turn to conclusion that Porous structure of hydroxyapatite-alumina composite has successfully produced using latex as space holder. Porosity characteristics of the sample show that density of the composite decreased with increased latex space holder. Increasing the percentage of latex space increase the amount of ash remaining that will fill the space between the grains formed by the burning of latex powder. Compressive strength results show that the addition of space holder weight fraction decreased the compressive strength of porous hydroxyapatite/$Al_2O_3$ composites. Increasing latex ash in the sintering process interferes diffusion process between HA particles. SEM observations of microstructure confirm that porous structure formed due to the process of burning down the space holder and an interconnected porous structure detected.

**Acknowledgment**

The authors would like to convey their great appreciation to Universitas Sriwijaya for supporting this research.

**References**

[1] D. Shi, D. Shi, and G. Jiang, "Introduction to Bioceramics," *Introd. to Biomater.*, pp. 3–12, 2010.
[2] T. Kanazawa, *Inorganic phosphate materials*, vol. 52. 1989.
[3] K. Usami and A. Okamoto, "Hydroxyapatite: Catalyst for a one-pot pentose formation," *Org. Biomol. Chem.*, vol. 15, no. 42, pp. 8888–8893, 2017.
[4] R. Choudhary *et al.*, 2016 "In-vitro bioactivity, biocompatibility and dissolution studies of diopside prepared from biowaste by using sol-gel combustion method," *Mater. Sci. Eng. C*, **68** pp. 89–100
[5] Z. Wu, Z. Zhou, and Y. Hong, 2019 "Isotropic freeze casting of through-porous hydroxyapatite ceramics," *J. Adv. Ceram.*, **8** pp. 256–264
[6] H. Yoshikawa, N. Tamai, T. Murase, and A. Myoui, 2009 "Interconnected porous hydroxyapatite ceramics for bone tissue engineering," *J. R. Soc. Interface*, **6** 3
[7] H. Noguchi, M. Sakane, A. Watanabe, T. Tsukanishi, Y. Wadano, and M. Yamazaki, 2014 "A novel unidirectional porous hydroxyapatite in canines," *Bioinspired, Biomim. Nanobiomaterials*, **3** pp. 228–234
[8] L. Stanev, M. Kolev, B. Drenchev, and L. Drenchev, 2016 "Open-Cell Metallic Porous Materials Obtained Through Space Holders—Part I: Production Methods. A Review," *J. Manuf. Sci. Eng.* **139** 050801

# Failure analysis of AISI 304 stainless steel pipeline transmission a petrochemical plant

**A Arifin[1], Gunawan[1,*], I Yani[1]**

[1] Department of Mechanical Engineering, Faculty of Engineering, Universitas Sriwijaya, Indralaya, 30662, Sumatera Selatan, Indonesia

* Email: gunawan@unsri.ac.id

**Abstract**. This paper presents the failure analysis of AISI304 stainless steel pipeline that was fabricated by welding and used for $CO_2$ transmission in a petrochemical plant. After a years of satisfactory operation, significant cracking was observed adjacent to the weld joints of valve and pipeline. The failure investigation was carried out on welded pipe samples using optical and fractography, micro hardness measurements, XRF and XRD testing. The hardness test results show faults occur in the HAZ region. Fractured surface macrostructure and morphology analysis shows inter-granular brittle fracture mode. The inter-granular fault begins with the presence of very intensive inter-granular corrosion, especially on the outer metal surface and progressively propagates to the inside of the metal. The results of XRD found peak of carbide from the secondary phase of SS 304 austenite steel. The material experienced sensitization that is indicated by the presence of carbides at the grain boundary.

**Keywords:** AISI-304; fractography; inter-granular cracks; Sensitization

## 1. Introduction

The selection of stainless steels for chemical transmission is based on mechanical properties, corrosion resistance, fabrication characteristics (such as weld ability), working environment, service temperature, and cost. However, corrosion resistance and mechanical properties are generally the most important factors whenever selecting a grade of stainless steel for a given application [1]. Among the many 300 series austenitic stainless steel grades, AISI-304 SS is widely used for the construction of large storage tanks for styrene monomers, because of its corrosion resistance in many environments and it's relatively low cost as compared to other 300 series steels [2].

The austenite phase in Austenitic Stainless Steel is stable even with very low nickel content, but with an increase in the percentage of manganese. However, this causes a decrease in the chromium solubility limit in the austenite phase. Therefore, the chromium content must be reduced to 15-16% by weight in Cr-Mn Austenitic Stainless Steel to keep the austenite structure intact. Cr content with a range of 18-20 Cr%, found on AISI 304 SS.

Ferritic and austenitic stainless steels are susceptible to intergranular corrosion when heated to temperatures between 400ºC and 900ºC [3]. The sensitization phenomenon occurs due to the combination of chromium with carbon which forms chromium carbide at the grain boundary. Increased carbide causes chromium grain boundaries that are susceptible to intergranular corrosion. This causes the appearance of microscopic cracks that will propagate at the grain boundary which causes the component to break even if it works below the yield point [4].

Valve (valve) is a component that has a very important role in ensuring a factory can operate. Failure to operate on the Valve will cause the factory to shut down. A valve is a device that regulates, directs or controls the flow of a fluid (gas, liquid, fluidized solids) by opening, closing, or closing part of the flow path. The valve can be operated manually, either by the handle, the lever pedal and others. Besides being able to be operated manually, the valve can also be operated automatically by using the principle of changing the flow of pressure, temperature, etc.

The problem of environmentally induced SCC and sensitization of stainless steel is of prime concerned to chemical, oil, & gas industries and refineries, and should be avoided, if practical, for any application. In general, poor welding practices promote the susceptibility of austenitic stainless steels to SCC. The main objective of this work is to analysis failure of AISI 304 stainless steel pipeline transmission.

## 2. Materials and Method

The investigation of the pipeline was made to determine the potential causes of the failure. A series experiment procedure performed to examine material properties after breaking. A fractographic based study of the fracture surface used to identify surface features by performing macro and micro structure analysis via both visually or microscopically that indicate the type and origin of failure. The present investigation also included hardness profile on the surface sample by using Vickers hardness tester VKH-2E, while composition analysed using XRF Analyser Niton XL2 and phase of material detected using X-Ray Diffraction Rigaku MiniFlex 600.

## 3. Results and Discussion

Composition testing was carried out to determine the type of material used using X-ray fluorescent (XRF) analyzers. The test is carried out in two different positions from the pipe, which is in the broken area and un-broken ends. The XRF results as can be seen in Table 1 show that the sample has a composition which is close to the SS 304 Austenitic element and there is no significant difference in chemical composition between the two test points. In the later position a small amount of molybdenum (Mo) is detected. However, the concentration of Mo elements is very low to influence the chemical properties of austenite steel, so the presence of this element can be ignored. In general, this element is added in an amount of at least 2% in austenite steel.

**Table 1**. XRF result of SS 304 Austenitic elemental analysis

| Element | Broken area | | Un-broken area | |
|---|---|---|---|---|
| | % | $\pm 2\sigma$ | % | $\pm 2\sigma$ |
| Cr | 18.19 | 0.62 | 18.47 | 0.56 |
| Mn | 0.928 | 0.459 | 1.07 | 0.41 |
| Fe | 70.96 | 1.16 | 70.97 | 1.06 |
| Ni | 9.79 | 0.75 | 9.31 | 0.67 |
| Mo | | | 0.178 | 0.069 |

The cross section of on failure pipe is shown in Figure 1.a Based on the visual analysis of macrostructure morphology on the fracture surface, the fracture mode shows brittle fracture. Brittle fractures on the morphology are clearly seen in quadrants I, II and IV. The fractured morphology also shows brown color that is believed to be due to corrosion (Figure 1.b). The presence of this corrosion also proves that the cracks exposed to the air for long time. Meanwhile in Figure 1.b also shows fractured surface due to fatigue as can be seen from the morphology marked by the presence of a beach mark in quadrant II.

The results of the hardness testing of fracture surface samples were carried out with the Vickers Method. The SS 304 material hardness is generally 189 VHN. As can be seen from figure 2 the Vickers value increases in longitudinal to the arrow direction compared to the hardness of material that is not exposed to welding heat. This increase occurs due to phase changes due to high exposure to heat and at the same time causes the brittle value of the material to increase.

(a) (b)

**Figure 1**. Fracture Morphology

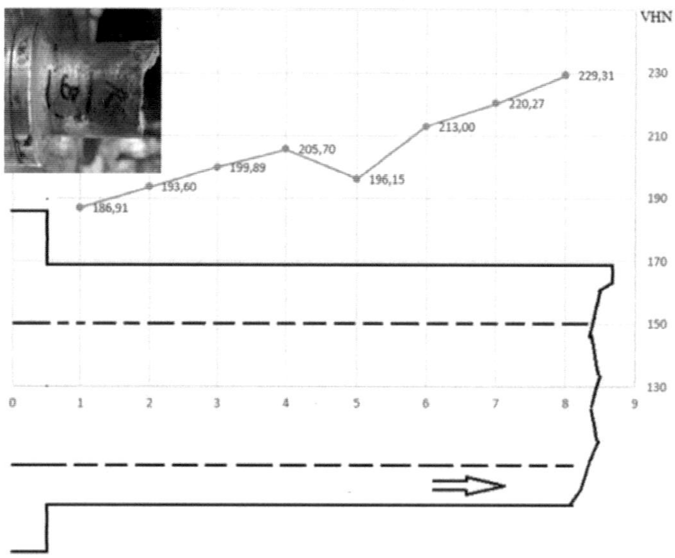


**Figure 2**. Vickers hardness test result

XRD testing is used for analysis of phase composition, compounds or material types as well as crystal characterization. The XRD result is depicted in figure 3 where material shows phase characteristics commonly found in austenitic steel namely phases of $\alpha$, $\gamma$ and Fe2O3. However, the results of XRD also found peak of carbide from the second phase (secondary phase) of SS 304 austenite steel. This phase usually occurs because the material has been exposed to high temperatures in a certain time range [5,6].

Observation of microstructure aims to analyze the microstructure present in the material can be seen at figure 4. Austenitic stainless steels do not respond to magnetic treatment without heat and are influenced by cold working. The steel microstructure consists of austenite grains which can show twinning. Twinning microstructure is detected in a number of places which are characteristic of austenite phase, this is in accordance with structures owned by SS 304 which have FCC structure. In this observation we also found the distribution of carbides in austenite and ferrite granules.

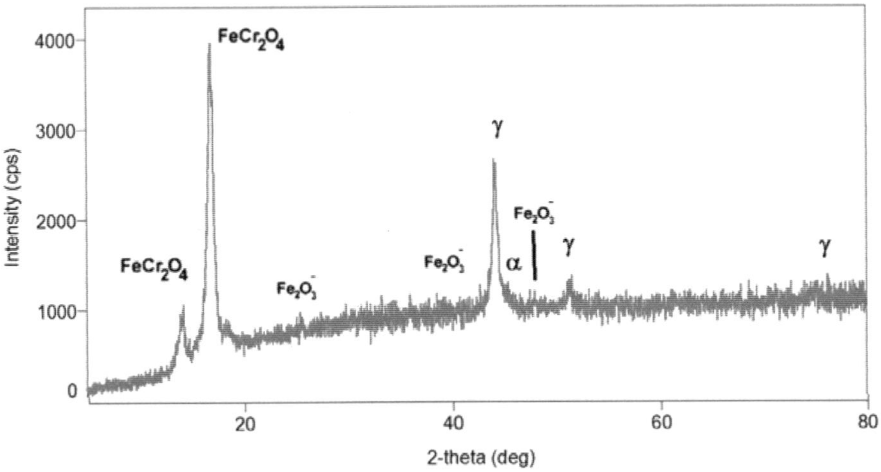


**Figure 3**. XRD test result

In addition, precipitation (carbide formation) is also found in grain boundaries. This precipitation occurs usually due to high temperatures that reach the range sensitization of SS 304 (600 - 850 ° C). If this happens the carbon will diffuse towards the grain boundary to form Cr Carbide. Cr carbide has brittle properties that are very detrimental to the mechanical properties of the material.

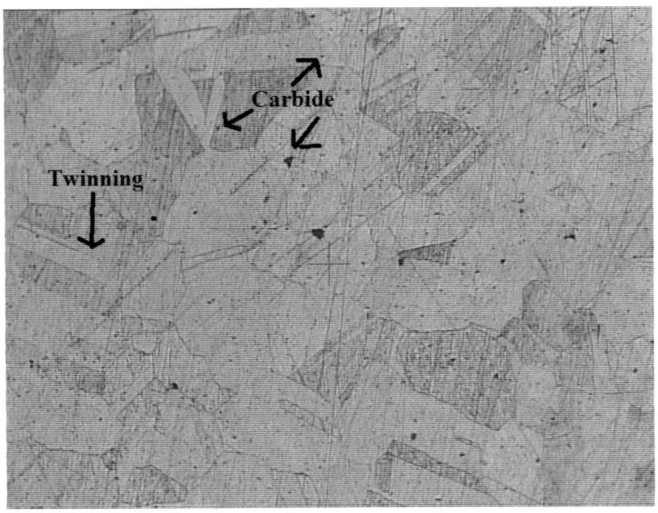


**Figure 4.** Microstructure observation

Stainless Steel SS 304 material Austenitic having a content of more than 0.05% C which is susceptible to inter-granular (weld decay) corrosion in the HAZ region. In addition, because of the high percentage of Cr in broken part, it is very vulnerable to the occurrence of the process of Cr carbide precipitation at the grain boundary, commonly known as sensitization. From the results of XRD testing it is known that the carbide phase Cr and carbide spread into grain and grain boundaries as seen in previous microstructure observations. In the temperature sensitization range carbon atoms will diffuse very quickly towards the boundary of the grain where they will join the Cr atom which forms Cr carbide.

SS 304 has a range of temperature sensitization between 600°C to 850°C. Visual observations obtained information that valve failure occurred on the HAZ. This confirms that failure can be ascertained due to precipitation of Cr carbide (sensitization). Phenomenon sensitization does not appear in the fusion zone because this position is the highest temperature peak. Sensitization will appear in the area closest to the fusion zone which has a lower peak temperature. This can be explained based on the thermal cycle of welding in figure 5.

Zone no. 2 is a zone where sensitization arises because it has a longer time range that allows precipitation. While zone 1 has a very short time range for precipitation and zone 2 peak temperatures are too low to reach zone of sensitization. Testing of hardness also proves that there is a significant increase in hardness towards the fault (weld) which proves that there is also a phase change in the material. Cr carbide itself has brittle properties.

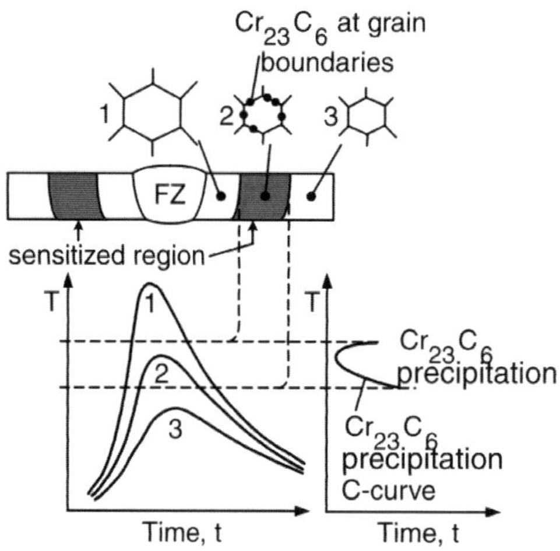


**Figure 5.** SS 304 sensitization zone

After experiencing sensitization, there will be Cr buildup at the grain boundary which causes a lack of Cr atoms in forming a layer of Cr oxide. This results in inter-granular corrosion which causes microscopic cracks to appear. The propagation of this crack is compounded by the vibrations received by the structure. Corrosion at this grain boundary starts from the metal surface then propagates towards the direction in the metal as shown in the figure 6.

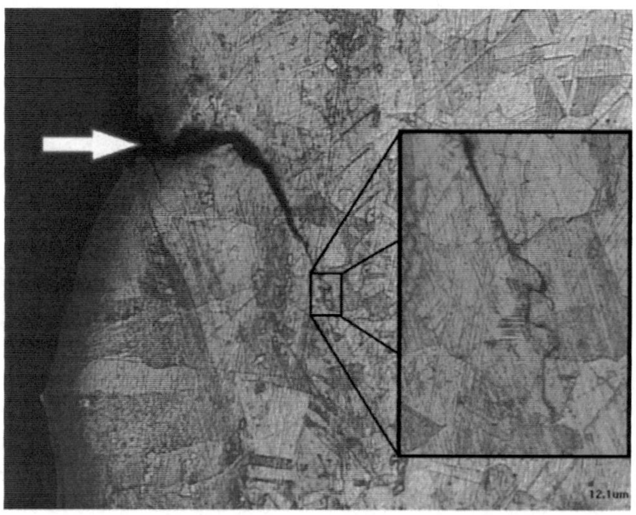

**Figure 6.** The propagation of inter-granular corrosion

## 4. Conclusion

XRF results indicate the material used is SS 304. The hardness test results show fracture occurs in the HAZ. Damaged surface macrostructure analysis and fault morphology shows inter-granular brittle fracture mode. The inter-granular fault begins with the presence of very intensive inter-granular corrosion, especially on the outer metal surface and progressively propagates to the inside of the metal SS 304 material experienced a sensitization indicated by the presence of carbides at the grain boundary.

**Acknowledgment**

The authors would like to convey their great appreciation to Universitas Sriwijaya for supporting this research.

## References

[1] -----2004, Handbook of Comparative World Steel Standards ASTM DS67B Third Edition John E. Bringas, Editor, ASTM International, West Conshohocken, PA.
[2] Kou S 2003 *Welding Metallurgy* (New York:John Wiley & Sons, Inc)
[3] Queiroz A A D 2012 AISI 304 Steel Intergranular Corrosion (Failure Analysis). In: 21th SAE BRASIL International Congress and Exhibition October, 2nd to 4, São Paulo, Brasil
[4] Callister W D and Rethwisch D G 2013 *Materials Science and Engineering: An Introduction, 9th Edition: Ninth Edition* (New York: John Wiley and Sons, Incorporated)
[5] https://www.phase-trans.msm.cam.ac.uk accessed at 22 September 2018
[6] Karimi, N., et al. 2008. "Characterization of the oxides formed at 1000°C on the AISI 304 stainless steel by X-ray diffraction and infrared spectroscopy." Applied Surface Science **254** 2292-2299.

# Development of aluminum matrix composite with hybrid reinforcement using stir casting route

**Gunawan[1], A Arifin[1,*], M Reza[1], A N P Wijaya[1]**

[1] Department of Mechanical Engineering, Faculty of Engineering, Universitas Sriwijaya, 30662 Indralaya, Sumatera Selatan, Indonesia

* Email: amir@unsri.ac.id

**Abstract.** Aluminum matrix composite (AMCs) has widely utilized in various industries such as automotive, aerospace maritime, electronics components, sports equipment. AMCs is considered as potential materials due to their excellent mechanical and physical properties. The aim of this work is to investigate effect of the addition SiC and fly ash on mechanical properties of AMCs. In this study, AMCs is fabricated using stir casting method with SiC (8 wt%) and Fly ash (0 wt%, 4 wt%, 8 wt% and 12 wt%) as hybrid reinforcement. Aluminum alloy was heated above its melting temperature in the steel crucible and poured into steel mold at 700°C with mixing speed of blade and mixing time are 350 rpm and 400 S, respectively. Mechanical properties of AMCs samples were characterized using tensile test, hardness test and impact test. Mechanical properties test reveals that AMCs with composition 12 wt% fly ash and 8 wt% SiC has maximum value in term of tensile test (15,859 kgf/mm2), impact test (10,22 Joule) and hardness test (57,15 BHN).

**Keywords:** Alumunium matrix composite, stir casting, fly ash

## 1. Introduction

Nowadays, development Metal Matrix Composites (MMCs) has been play important role in material innovations. Aluminium matrix composite (AMC) is classified as MMCs refer to aluminium alloy as matrix used. AMC has some potential properties for engineering structure application such as low density, stiffness, wear resistance, fatigue resistance, high thermal conductivity, high energy absorption and better stability at elevated temperature. Owing to this advantages, AMCs are utilized widely for various advanced applications such as automotive part aerospace, marine and nuclear industries. In weight reduction issue, automotive industries are being subjected to reducing strict of pollution and fuel consumption, AMCs have to be alternative material to change conventional materials [1, 2].

Ceramic particles generally used as reinforcement in AMCs fabrication. The addition of ceramics particles has significant contribution to increase hardness, stiffness and strength of composite [3, 4]. Silicon carbide and alumina are ceramic particles which has been ordinarily used as reinforced [5, 6]. AMCs reinforcement materials in term of particle shape can be divided into four categories; continuous fibers, short fibers, whiskers and particles. Continuous fibers is effective to strengthening in certain direction meanwhile reinforcement particles more attractive due low cost and flexibility.

Commonly fabrication of AMCs can be divided into two methods, solid state processing and liquid state processing. Stir casting can be classified as liquid state processing with some advantages such as lower cost, near net shape and mass production. AMCs Fabrication methods has important role to determine mechanical and physical properties of composite. Moreover, it is having also give contribution to determine of the interfacial bonding between reinforcement material and matrix.

AMCs Fabrication still faces several challenging issues to solve such as distribution of reinforcement and bonding of ceramic particles with aluminum alloy as matrix. At elevated temperature during casting process promotes chemical reaction which leads degradation of ceramic particles [7, 8]. After occurs degradation on ceramic particles, expectation mechanical properties of composite is difficult to obtain. Fly Ash is part of the remaining coal combustion consisting of amorphous fine particles and ash which is an inorganic material formed from mineral materials.

Stir casting is one of the routes to produce aluminum matrix composites in liquid conditions using a stirrer, which is carried out above the melting temperature of the metal. The advantage of the stir casting method is simple process, flexible and can be used for products in large quantities. Moreover, it is the most economical for producing aluminum matrix composites and allows to be used for large component fabrication processes. The main objective of this study is to analyze the effect of additional of SiC and fly as as hybrid reinforcement in Aluminum matrix composite.

## 2. Materials and methods

In this study, fly ash was collected at surrounding area at one power plant in Tanjung Enim, South Sumatera, Indonesia. Fly ash particles in this work has irregular shape with 40 mesh size. Chemical composition was performed to investigate chemical component of fly ash. Fly ash consist of some ceramics component as can be seen on the Table 1.

**Table 1**. Chemical composition of fly ash.

| components | $SiO_2$ | $Al_2O_3$ | $Fe_2O_3$ | $TiO_2$ | CaO | MgO | $K_2O$ | others |
|---|---|---|---|---|---|---|---|---|
| Wt % | 57.95 | 28.15 | 4.75 | 0.93 | 3.74 | 1.73 | 0.62 | 2.13 |

Homogen distribution of reinforce particles in aluminum matrix composite is critical issue which has responsible for mechanical and physical properties of AMCs. Processing parameters such as casting temperature, mixing speed and type of blade are some parameters should be considered [1]. In this work, Electric motor with control speed was utilized to generate carbon steel stirring blade. Furnace temperature to melt aluminum alloy was regulated using electric blower to adjust air flow in to burning chamber. The molten aluminum alloy was stirred until generate vortex, during molten aluminum alloy was stirred preheat fly ash was added slowly. Furthermore, at temperature 700°C molten aluminum alloy was poured into carbon steel mold.

AMCs specimens were prepared through machining process for tensile, hardness and impact strength. Universal testing machine BH-3CF RAT-30 30tf was conducted to investigate tensile strength refers to the JIS Z 2201 standard. Meanwhile impact strength was investigate using Charpy Impact testing machine based on JIS Z 2202 Standard.

## 3. Results and discussion

Figure 1 shows the AMCs specimen for tensile test and impact test that fabricated using stir casting after through machining process.

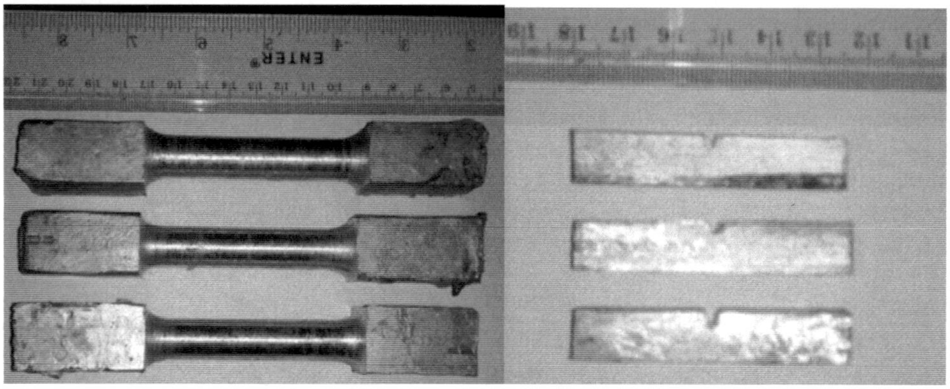

**Figure 1**. Tensile and impact test sample with composition 4 wt% Fly ash and 8 wt% SiC

Tensile strength is important properties that considered primary properties for engineering design. Figure 2 shows tensile strength result of AMCs specimen with various composition. Maximum tensile strength value (15.859 kgf/mm2) was obtained in composition 12 wt% fly ash and 8 wt% SiC, meanwhile minimum tensile strengths in AMCs Specimen without composition of fly ash.

The increasing of amount of fly ash percentage has contribution to enhance tensile strength value of AMCs. On the other hand, the increasing of fly ash has effect to decrease elongation of AMCs. The addition of fly ash percentage with existence of SiC has effect to reduce ductility of AMCs as shown measurement of elongation as shown on Figure 3.

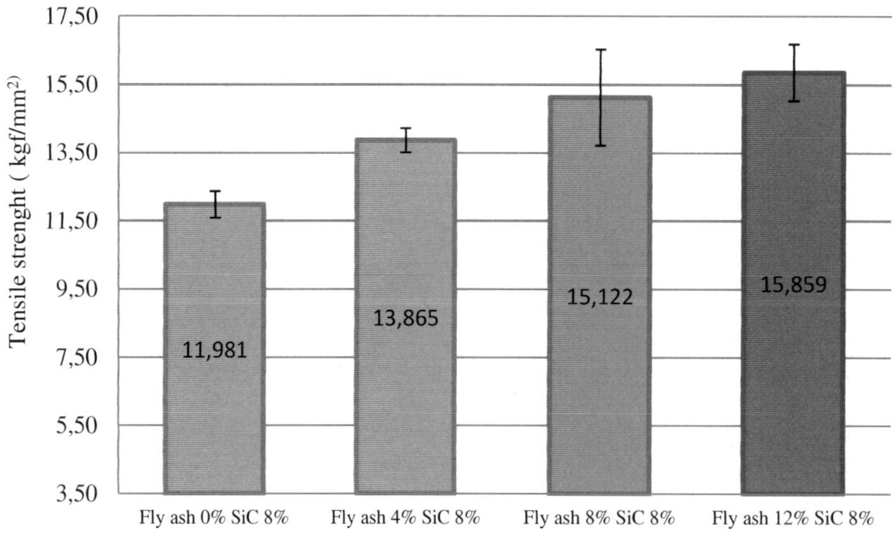


**Figure 2**. Tensile strength of AMCs with 8 wt% SiC and various weight percentage of fly ash content

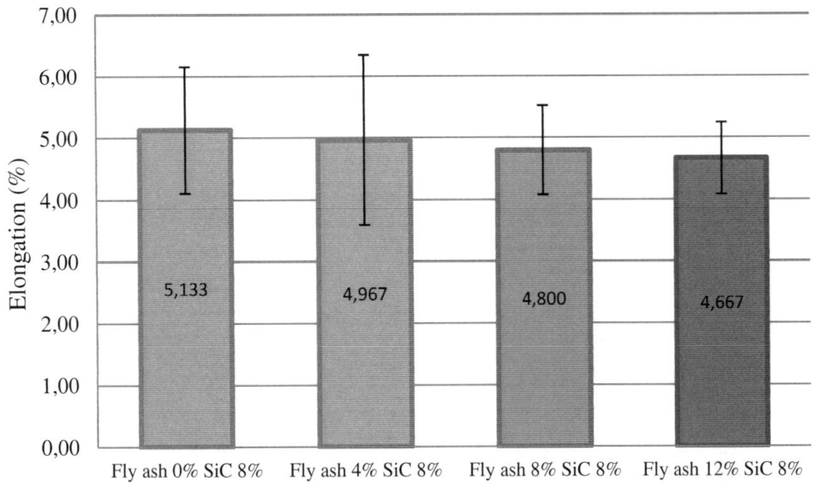


**Figure 3.** Elongation of AMCs with 8wt% SiC and various weight percentage of fly ash content

The impact tests for AMCs sample were conducted using Charphy test. Impact tests result show that AMCs obtain maximum value on the composition 12 wt% fly ash and 8 wt% SiC in the other word that the sample absorb more energy than other composition.

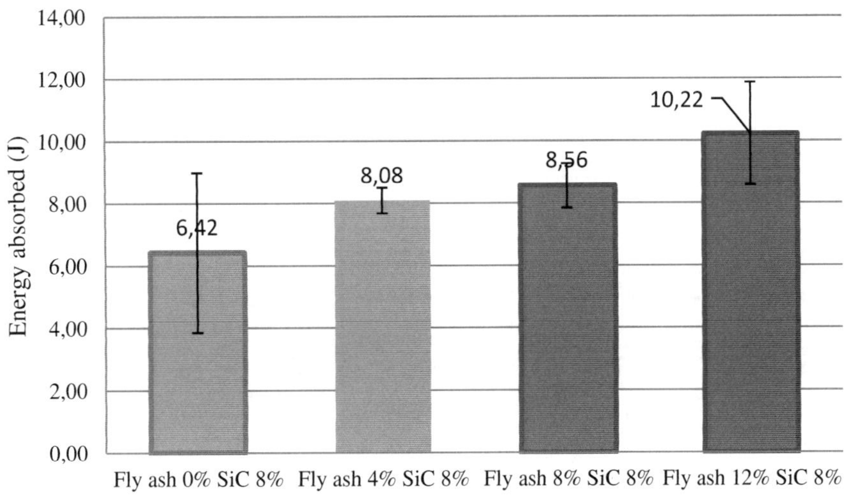


**Figure 4** Energy absorbed of AMCs with 8wt% SiC and various weight percentage of fly ash content

Hardness properties of AMCs was characterized using Brinnel hardness tester type BH-3CF with Indenter ball (5 mm). The load (500 kgf) was subjected to the AMCs sample refer to JIS B 7724 standard. Brinell method is chosen due to more accurate to analyses heterogenic material such as metal matrix composite. Distribution of reinforcement materials is problem encountered in fabrication metal matrix composite. It always occurs due to the density different between matrix alloys and reinforcement materials.

The position of reinforcement particles in matrix composite are determined by during mixing, after mixing and after solidification. Hardness measurement result revealed that hardness value of AMCs sample increase with the additional of fly ash percentage. Maximum hardness value (57.15 BHN) was obtained in composition 12 wt% fly ash and 8 wt% SiC. Otherwise, minimum hardness value (50.61 BHN) was obtained in composition 8 wt% SiC without fly ash content. The increasing of hardness value is determined by the additional of fly ash percentage with has some ceramic components with high hardness value. When hardness test was carry out, indenter ball was resisted by reinforcement materials.

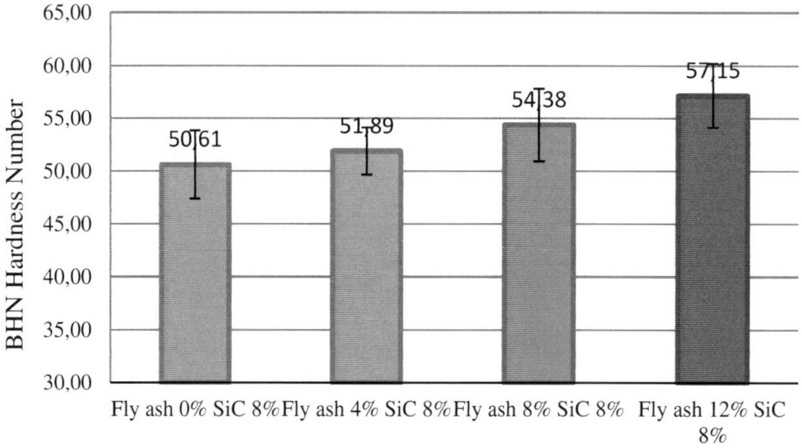


**Figure 5** Hardness test result of AMCs with 8wt% SiC and various weight percentage of fly ash content

The Ideal casting process for fabrication aluminum matrix composite if distribution of reinforcement particles uniformly dispersed throughout the matrix. Distribution of reinforcement particles is complicated problem there are many factors to determine as previous discuss. Moreover, not only from processing variables but also from size, shape and physical properties of reinforcement particle and matrix have strong contribution. To investigate microstructure of AMCs and distribution of reinforcement particles.

Optical microscope was utilized to analyzed morphology of AMCs as shown on the Figure 6. Based on the optical microscope investigation that reinforcement particles have been found particles dispersed on the surface of AMCs.

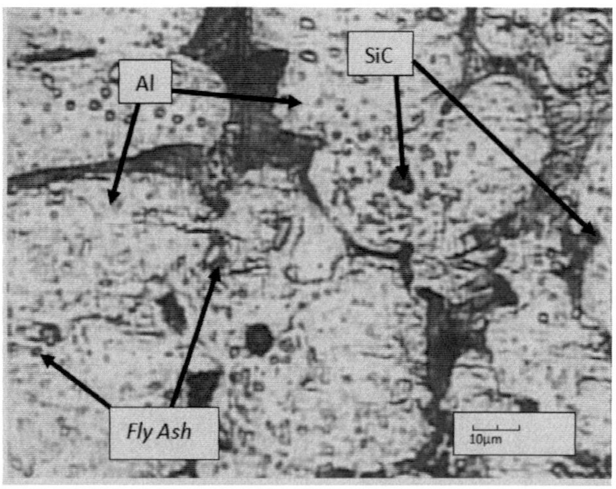


**Figure 6** Microstructure aluminum matrix composite with 4 wt% *fly ash* and 8% $S_iC$

## 4. Conclusions

Fabrication of Aluminum matrix composite with hybrid reinforced (SiC + fly ash) using stir casting method was successfully performed. In this study the addition of various fly ash (0, 4, 8 and 12 wt %) has contribution to improve mechanical properties of AMCs such as tensile strength, energy absorption and hardness. On the other hand, Elongation of AMCs tend to decrease with the additional of percentage of fly ash. Maximum value is obtain for tensile test, impact and hardness AMCs composition with 12 wt% *fly ash* and 8 wt% SiC are 15,859 kgf/mm$^2$, 10,22 Joule and 57,15BHN, respectively.

## References

[1] Macke, A., B.F. Schultz, and P. Rohatgi, 2012 Metal Matrix Composites Offer the Automotive Industry an Opportunity to Reduce Vehicle Weight, Improve Performance. Advanced Materials & Processes. March.
[2] Joost, W.J. and P.E. Krajewski, 2017 Towards magnesium alloys for high-volume automotive applications. Scripta Materialia. **128**: p. 107-112.
[3] Singh, J. and A. Chauhan, 2016 Characterization of hybrid aluminum matrix composites for advanced applications – A review. *J. of Materials Research and Technology*. **5**(2): p. 159-169.
[4] Kulkarni, S.G., J.V. Meghnani, and A. Lal, 2014 Effect of Fly Ash Hybrid Reinforcement on Mechanical Property and Density of Aluminium 356 Alloy. *J. Procedia Materials Science*,. **5**: p. 746-754.
[5] Arifin, A., et al., 2018 Effect of Fly Ash As Reinforcement on Mechanical Properties of Aluminum Scrap Based Hybrid Composite. *J. of Engineering Science and Technology*, **13**(10): p. 3080 - 3091.
[6] Bodukuri, A.K., et al., 2016 Fabrication of Al–SiC–B4C metal matrix composite by powder metallurgy technique and evaluating mechanical properties. Perspectives in Science, **8**: p. 428-431.
[7] Noble, B., A.J. Trowsdale, and S.J. Harris, 1997 Low-temperature interface reaction in aluminium–silicon carbide particulate composites produced by mechanical alloying. *J. of Materials Science*, **32**(22): p. 5969-5978.
[8] Braaten, O., A. Kjekshus, and H. Kvande, 2000 The possible reduction of alumina to aluminum using hydrogen. Jom, **52**(2): p. 47-53

# Effect of fly ash content in Aluminum matrix composite through stir casting method on mechanical and physical properties

**Gunawan[1], A Arifin[1,*], M A Akbar[1], I Asura[1]**

[1] Department of Mechanical Engineering, Sriwijaya University, 30662 Indralaya, Sumatera Selatan, Indonesia

[*] E-mail: amir@unsri.ac.id

**Abstract** Metal Matrix Composite (MMC) is a potential innovation material to meet the requirements required in various applications. such as in the aerospace, automotive and mining sectors. Stir casting has been known widely as a methode to produce MMC. In this study, Alumunium alloy and fly ash is utilized as matrix composite and reinforced phasing, respectively. Effect for mechanical and physical properties are investigated for the addition fly ash (4, 8 and 12wt%). Alumunium alloy melt with fly ash is stirred with 350 rpm for 5 minutes and poured into the mold at temperature of 800°C. Hardness, impact and density test are performed to investigate the characterization of specimen. The experimental results show that maximum value is obtained at 8 wt% fly ash with 65.83 HB, 11. 8 Joule, 14.299 Kgf/mm$^2$ and 2.7 g/cm$^3$ for hardness, impack, tensile strenght and density, respectively.

**Keywords**: Aluminum, fly ash, stir casting, composite, mechanical properties

## 1. Introduction

Aluminum alloys has been widely utilized in various application such as aerospace, automotive and mining process [1-3]. Aluminum alloy is potential material owing mechanical and physical such as strength, density and corrosion resistance [2]. Occasionally, special part of structure running in extreme condition such as high temperature, cyclic loading and under impact loading. In those condition aluminum alloy component is required to improve mechanical properties.

Fly ash is material residual obtained from coal combustion which is classic problem surrounding power plant. Fly ash has a density of 1 g / cm$^3$ and comprising to $SiO_2$, $Al_2O_3$, $Fe_2O_3$ and oxide content such as Mg, Ca. The chemical, mechanical properties, morphology and size of fly ash is potential material to replace classic materials to produce glass and ceramic part. In term of environmental and ecological problem, the utilization on fly ash as raw material is expected has contribution reduce waste that is generated by power plants.

MMC (Metal Matrix Composite) is consist of metal as matrix and reinforcement particle. In term of Aluminum alloys as matrix, commonly known as Aluminum Matrix Composite (AMC). AMC is a type of Metal Composite Material with aluminum alloys as the matrix and ceramic materials such as $SiO_2$, $Al_2O_3$ as reinforcement particles [4, 5]. AMC has a prospective development that is promising, based on its good characteristics, such as high hardness and strength.

Commonly, Fabrication of MMC can be classified into liquid and solid state processing methods. Liquid processing method comprising to squeeze casting, compo casting, stir casting and infiltration method while solid state processing techniques consist of; friction stir processing, diffusion bonding, spray deposition. Stir casting method is most valuable because of its simplicity and applicable to mass production. Moreover, stir casting as common casting method also has applicable to produce near net shape part.

Some parameter processes play important role during stir casting conducted such as stirring time, pouring temperature, mixing time, speed of blade and reinforcement particle composition [6, 7]. Previous works reported by some authors that composition of SiC and Fly ash has the most significant contribution to determine of hardness properties [5, 8]. Some efforts have been conducted to improve mechanical and physical properties of aluminum alloy through stir casting route [5, 6, 8-11]. The increasing of percentage of fly ash in aluminum alloys (Al/3Cu/8.5Si) has effects to increase the strength properties. On the other hand, some properties such as ductility and impact strength tend to decrease as reported by the other author [12].

The objective of this research is to produce Aluminum matrix composite with fly ash as reinforcement particle through stir casting route. The addition of percentage of fly ash is investigated in term of mechanical properties effect of AMC. The microstructure and mechanical properties of the fabricated AMC were studied.

## 2. Experimental procedure

In this work, Aluminum matrix composite (AMC) was fabricated with help of a stir casting process. Stir casting route is method of melting the material with continuous mixing using a blade and directly pouring the melt into the cavity mold. Stirring process is conducted during the process to avoid agglomeration reinforcement particles. It is also necessary factor to improve homogeneous distribution of reinforcement particles in matrix [13].

Stir casting were performed with parameter process mixing speed, mixing time and mixing temperature for 350 rpm, 5 minute, 660°C and respectively. Aluminum alloy was melted in the crucible that heated to melting temperature. Fly ash was added externally in a fixed proportion (4, 8 and 12 wt%) into aluminum alloy melt. Magnesium alloy with 2.5wt% was added to aluminum alloy composite to increase wettability between aluminum alloy and reinforcement particle. Table 1 shows aluminum alloys composition was used as a matrix in this work with fly ash as reinforcement particles.

**Table 1** Chemical composition of Aaluminium alloy (wt%)

| Element | Fe | Mn | Cu | Zn | Pb | Al |
|---|---|---|---|---|---|---|
| wt% | 0.424 | 0.649 | 0.345 | 0.286 | 0.033 | balance |

Mechanical properties of AMC cast part was characterized using Brinnel Hardness Testing Machine using JIZ Z2243 standard for measuring hardness value. The ability of AMC specimens to absorb energy until fracture were measured using impact test base on JIZ Z2202 standard. Moreover, tensile strength of AMC specimen was investigated using Universal Testing Machines based on JIS Z 2201 standard. Scanning electron microscopy (SEM) was used to study the surface morphology cast part of AMC.

## 3. Results and Discussions

Reinforcement particle distribution is main factor to determine performance of mechanical and physical properties of AMC [13]. Impact test was performed to investigate of toughness properties of aluminum matrix composite. Toughness properties in term of mechanical definition is the ability of a material to absorb energy up to fracture. Figure 1 shows impact test results of AMC cast part for various weight percentage of fly ash. The addition of fly ash has effect to influence impact properties of AMC. Maximum hardness value (11.879 Joule) was obtain in composition 8wt% of fly ash.

On the other hand, composition 4wt% and 12wt% were not significant to change impact properties compare to specimens without fly ash content.

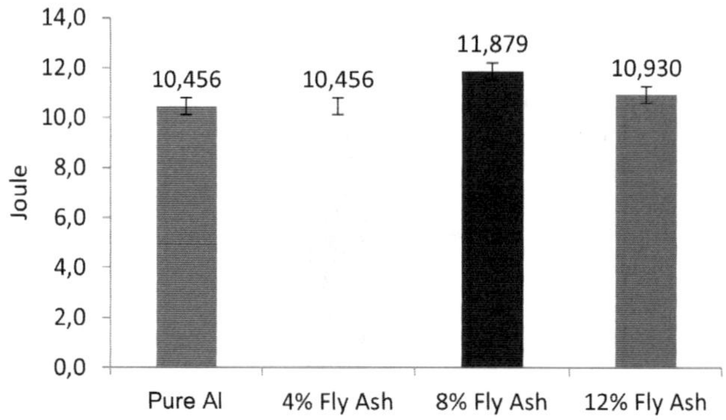


**Figure 1** Impact test results of aluminium matrix composite for various percentage of fly ash

Figure 2 shows Effect of various percentage fly ash on the tensile strength of the Aluminum matrix composite. Compare to aluminum alloy without reinforcement particles, the addition of reinforcement particle into aluminum alloy can improve tensile strength of AMC. The tensile strength tends to increase with the addition of fly ash percentage from 4 wt% to 8 wt%. The maximum tensile strength value (11.879 kgf/mm2) was achieved at 8 wt% and minimum tensile strength at 12 wt% (13.443 kgf/mm2). The reduction of tensile strength value after 8 wt% fly ashes due to excessive of reinforcement particles in AMC which is refer to in term of the ability matrix to bond reinforcement particles.

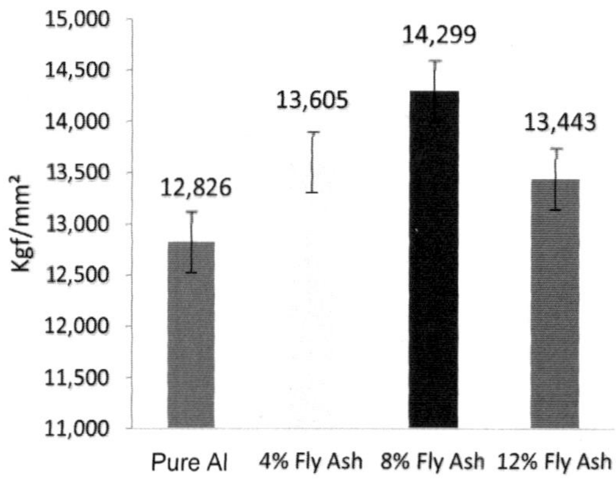


**Figure 2** Effect of various percentage fly ash on tensile strength of Aluminum matrix composite

Density measurement of aluminum matrix composite samples were conducted using Archimedes method. Figure 3 shows density result of aluminum matrix composite with various percentage of additional of fly ash. Maximum density value was obtained in fly ash composition 8 wt% on the other hand minimum density value of AMC in composition 4 wt% for 2.716 g/cm$^3$ and 2.684 g/cm$^3$, respectively. The additional of fly ash content more than 8 wt% has play important role to reduce density value of AMC. The occurrence of porosity on the AMC specimen is believe to be main factor to reduce density value.

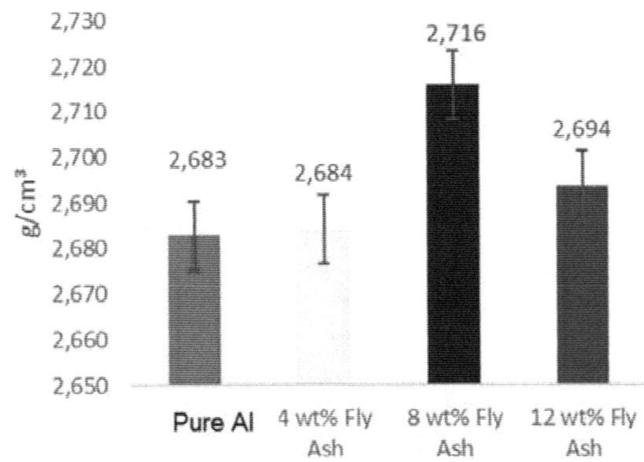


**Figure 3** density of Aluminium matrix composite with various percentage of fly ash

The porosity percentage in the specimen causes the material to be light and not dense. When tensile testing is carried out, porosity will form a large hole so that crack propagation occurs which causes a break in the specimen. This phenomenon makes the value of tensile strength in addition 4 and 12% become brittle.

The addition of 8 wt% fly ash has a maximum tensile strength value due to the bond between matrix and amplifier is still optimum. The value of high tensile strength at the addition of 8 wt% fly ashes can also be seen from the density value. The density value at the addition of 8% fly ash also has the highest value compared to pure aluminum and aluminum with the addition of 4 and 12% fly ash. Maximum density value is obtained at the addition of 8% of fly ash as an indication low porosity level compare to the other sample.

The presence of porosity in aluminum matrix composite detriment to mechanical properties especially tensile strength. Porosity in a material under tensile load is a significant contribution to accelerating crack propagation in the material. The occurrence of porosities on stir casting product due to chemical reaction between reinforcement particle (fly ash) and aluminum alloy [14]. Moreover, stirring process also give contribution for bubbles formation during stirring was conducted.

Morphology surface of aluminum matrix composite was investigated using scanning electron microscopic. Fracture surface morphology of sample in cross section area was observed using SEM analysis to determine fracture mode of material. Figure 4 shows SEM analysis for surface morphology of AMC cross section. SEM observation revealed that brittle fracture mode was observed which is indicating the material tend to brittle. Moreover, formation some dimples also was founded in the morphology AMC. Furthermore, analysis SEM shows the existence of porosities occur in some places in the surface which has significant contribution to detriment mechanical properties of AMC.

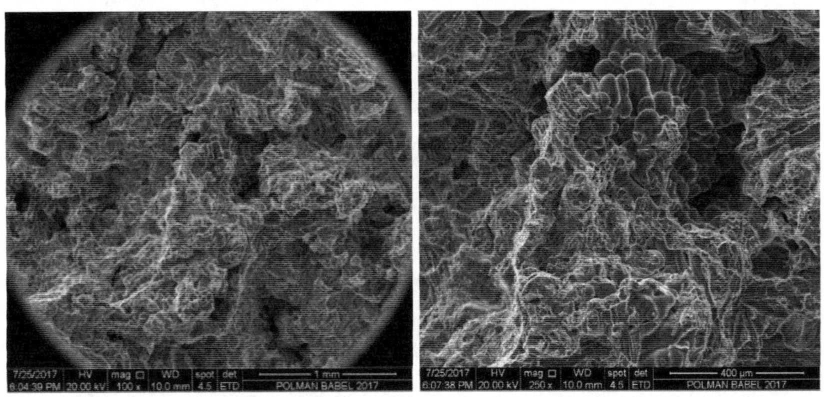

**Figure 4** SEM images analysis of aluminum matrix composites

Mechanical properties test comprising to tensile, impact and density test revealed that the maximum value were obtained for fly ash composition 8 wt%. At This composition alumunim alloy as matrix and fly ash as reinforcment phasing in best condition to generate optimum value of mechanical properties.

## 4. Conclusion

Fabrication of aluminum matrix composite with fly ash as reinforcement particles was successfully fabricated through stir casting method. The addition of fly ash percentage has play important role to determine mechanical properties of AMC. The composition 8 wt% of fly ash content shows maximum value in term impact, tensile and density measurement. SEM analysis indicates the composites indicates the distribution of the fly ash components throughout the aluminum matrix. Furthermore, there are some porosities were observed in cross section area. Porosity is believing to be detrimental factor to reduce mechanical properties of AMC product. It was supported by density results which is the AMC specimen with higher value of density has maximum value of impact and tensile test. The addition of fly ash into aluminum alloy melt has effect to reduce ductility of composite base on surface morphology of specimen.

## References

[1] Mavhungu S T, Akinlabi E T, Onitiri M A and Varachia F M 2017 Aluminum Matrix Composites for Industrial Use: Advances and Trends *Procedia Manufacturing* 7 178-82

[2] Hirsch J 2014 Recent development in aluminium for automotive applications *Transactions of Nonferrous Metals Society of China* **24** 1995-2002

[3] Efzan M N E, Syazwani N S and Abdullah M M A B 2016 Microstructure and Mechanical Properties of Fly Ash Particulate Reinforced in LM6 for Energy Enhancement in Automotive Applications *IOP Conference Series: Materials Science and Engineering* **133** 012046

[4] Wang W, Li Q, Ma R, Cui Y, Cao X and Wang X 2017 Effects of the Si contents of an infiltrating aluminium alloy on the microstructure and strength of SiC matrix composites *Ceramics International* **43** 12526-33

[5] Ezatpour H R, Sajjadi S A, Sabzevar M H and Huang Y 2014 Investigation of microstructure and mechanical properties of Al6061-nanocomposite fabricated by stir casting *Materials & Design* **55** 921-8

[6] Soltani S, Azari Khosroshahi R, Taherzadeh Mousavian R, Jiang Z-Y, Fadavi Boostani A and Brabazon D 2015 Stir casting process for manufacture of Al–SiC composites *Rare Metals* **36** 581-90

[7] Sun Y, Lyu Y, Jiang A and Zhao J 2014 Fabrication and characterization of aluminum matrix fly ash cenosphere composites using different stir casting routes *J Mater Res* **29** 260-6

[8] Arifin A, Gunawan G, Yani I, Yanis M and Pradifta R 2017 Optimization of Stir Casting Method of Aluminum Matrix Composite (AMC) for the Hardness Properties by Using Taguchi Method *Jurnal Kejuruteraan* **29** 35-9

[9] Balasubramanian I and Maheswaran R 2015 Effect of inclusion of SiC particulates on the mechanical resistance behaviour of stir-cast AA6063/SiC composites *Materials & Design (1980-2015)* **65** 511-20

[10] Sahin Y 2003 Preparation and some properties of SiC particle reinforced aluminium alloy composites *Materials & Design* **24** 671-9

[11] Moses J J, Dinaharan I and Sekhar S J 2014 Characterization of Silicon Carbide Particulate Reinforced AA6061 Aluminum Alloy Composites Produced via Stir Casting *Procedia Materials Science* **5** 106-12

[12] Kumar K R, Mohanasundaram K M, Subramanian R and Anandavel B 2014 Influence of fly ash particles on tensile and impact behaviour of aluminium (Al/3Cu/8.5Si) metal matrix composites *Sci Eng Compos Mater* **21**

[13] Sharma V K, Singh R C and Chaudhary R 2017 Effect of flyash particles with aluminium melt on the wear of aluminium metal matrix composites *Engineering Science and Technology, an International Journal* **20** 1318-23

[14] Fan L-J and Juang S H 2016 Reaction effect of fly ash with Al–3Mg melt on the microstructure and hardness of aluminum matrix composites *Materials & Design* **89** 941-9.

# Study on genetic algorithm (GA) approaches for solving Flow Shop Scheduling Problem (FSSP)

**A Syarif [1,*], W Wamiliana [2], P Lumbanraja [1] and M Gen[3,**]**

[1] Department of Computer Science, Faculty of Mathematics and Sciences, University of Lampung, Bandar Lampung, Indonesia
[2] Department of Mathematics, Faculty of Mathematics and Sciences, University of Lampung, Bandar Lampung, Indonesia
[3] Research Institute for Science and Technology, Tokyo University of Science (TUS), Tokyo, Japan

* Email: admi.syarif@fmipa.unila.ac.id

** Email: mitsuogen@gmail.com

**Abstract.** The scheduling problem is known as one of the well-known optimization problems. It occurs in many situations of our daily-life applications, especially in industrial fields. One type of scheduling problem is called Flow Shop Scheduling Problem (FSSP). It belongs to the class of NP-complete problem. During the last decades, researches on exploring more accurate and efficient heuristic methods to solve hard optimization problems have taken considerable attention from researchers. Among them, GA has been one of the powerful and widely used algorithms. In this paper, we present two GA approaches to solve FSSP. The main objective is to investigate the effectiveness and the efficiency of GA based on different variations of the chromosome representation, referred to as the job-based GA (jb-GA) and machine-based GA (mb-GA). We conducted numerical experiments using standard test problems (Benchmark test problems). We also present the comparison of results with those given by another heuristic algorithm (NBH Algorithm) and the optimal solutions reported in the literature. Those demonstrate the jb-GA is more effective and efficient almost all of the time. The current limitation of this approach, like many other heuristic methods, is that it still sometimes gives the near-optimal solutions.

**Keyword:** genetic algorithm, scheduling, benchmark, flow shop

## 1. Introduction

The scheduling problem concerns the allocation of limited sources over time to perform the task to satisfy specific criteria. This problem exists everywhere in our daily life, especially in industrial applications. However, it is known as one of the hard combinatorial optimization problems. It has highly complex constraints and belongs to the class of NP-hard problems. Despite its NP-Hardness and its importance, during the last decades, many solution methods have also been proposed to solve it [1]. There have been many variations of scheduling problems for different applications [2,3]. In general, there are two types of scheduling problems discussed in the literature. Those are Flowshop Scheduling Problem (FSSP) and Jobshop Scheduling Problem (JSSP).

FSSP occurs when $m$ machine process $n$ jobs in the same sequence. A different series usually will differ in terms of processing time. An example of FSSP occurs in manufacturing facilities where jobs moved from machine-to-machine. It is widely known as one of the NP-complete optimization problems with n! Possible schedule. Recently, there have been many variations of FSSPs intensively studied in the literature for various applications. There have many variants of solution methods to solve FSSP; most of them are heuristic methods [4].

Nowadays, as computers rapidly increased, researchers have more attention on applying heuristic methods such as Genetic Algorithm (GA), Tabu Search (TS), and Simulated Annealing (SA) for solving various NP-hard/NP-complete optimization problems including Scheduling Problem. Most of the objective is to develop both accurate and efficient heuristic methods. Among them, GA is the most powerful and widely used [5]. Several researchers reported the successful implementation of GA to solve a wide variety of real-world applications, including engineering, economics, finance, manufacturing, agriculture, business, etcetera. In our previous works, we also have reported the success of GA for various combinatorial optimization problems [6], [7], [8], and [9]. Though GA has been a versatile approach for searching the global optimality, it also has a disheartening weakness in gaining too many time to reach optimal solutions. The success of GA depends on several factors, including an efficient design of the chromosome representation, method of crossover and mutation, selection methods, and the value of GA parameters. Thus, research on determining an efficient design of the GA approach for a specific problem becomes very crucial.

In this research work, we present two GA approaches called job-based GA (jb-GA) and machine-based GA (mb-GA) to solve FSSP. These approaches differ in the way to represent the chromosome. Our primary intention is to investigate the effectiveness and efficiency of GAs to solve FSSP. We carried out some numerical experiments using Benchmark test problems to see the performances of the algorithms [10]. The results are comparable with those given by another algorithm called the NBH algorithm [11].

We organize the remainder of this paper as follows: In Section 2, we give a brief overview of the FSSP. Section 3 describes the design strategies of the proposed GA approaches, including the design of chromosome representations, genetic operations, and selection strategy. We present the numerical experiment results and the comparison with other heuristic methods in Section 4. Finally, Section 5 describes the conclusion showing the remarkable effectiveness of the approaches.

## 2. Flow Shop Scheduling Problem

Flow shop scheduling is one of the problems that s follows: There is a set of $m$ number machines and $n$ number of jobs. Each job consists of $m$ operation(s) that must be processed with a different device. The sequence for processing all jobs in the $m$ machine(s) is the same. $t_{ij}$ ($i$ =1, …, $n$; $j$ =1, …, $m$) denote the processing time of job $i$ by using machine $j$. For FSSP, we have the following assumption:

- Every job has to be processed on all machines in the order $j$ = 1, 2, …., $m$.
- Each machine processes just one job at a time.
- Operations are not preemptive.
- The processing times include Set-up times for the operations.
- Operation sequences of the jobs are the same on every machine.

The usual objective function is to determine the schedule (the processing job sequence on the machine or machine sequence to process jobs) with minimum Makespan. There are also some different objective functions used, i.e., total tardiness, mean flowtime, and so on. We can find the mathematical formulation of FSSP in [12].

## 3. Design of The GAs

In this section, we describe the GA, which is one of the accurate and efficient heuristic methods to solve hard optimization problems. It was first introduced by Holland [13] and popularised by several researchers, including [14], [15], and [4].

### *3.1. Initialization*

When implementing GA for a specific application, the first step is to find a way to represent the possible problem solution. Here, we applied the permutation-based representation called job-based GA (jb-GA) and machine-based representation (mb-GA). For jb-GA, a list of $n$ jobs represents the chromosome. Each job appears once in the list. Thus, this representation will always yield a feasible schedule. The order represents the sequence of the job processed in each machine. We construct the schedule following the order in the list.

Similarly, for mb-GA, the chromosome represents the order of the machine to process each job. It will be a list of $m$ machines. We generate the value of each gene in the chromosome randomly. As an example, Figure 1 illustrates a chromosome for the problem ta001 (20 jobs and five machines).

| | | | | |
|---|---|---|---|---|
| 1 | 5 | 3 | 2 | 4 |

**Figure 1.** An example of chromosome representation

### *3.2. Crossover and Mutation*

The purpose of crossover operation is to make replication of the chromosome. It has a vital role in the success of GA. For permutation-based representation, we cannot use simple two-point crossover operations. There are many variants of crossover operations usually used for permutation representation, such as Partially Matched Crossover (PMX), Position-based crossover (PX), and Weight mapping crossover (WMX) [4]. Here, we adopt the PMX crossover as follows:

**Procedure:** **PMX**
**Step 1**: Select a section of chromosome randomly
**Step 2**: Exchanged each substring
**Step 3**: Determine the mapping of genes in each substring
**Step 4**: Update the chromosome with information on Step. 3

Another essential feature of GA is the mutation operation. It is usually done by exchanging the data within a chromosome to prevent premature loss of information. In this paper, we adopt the inversion mutation that selects two positions within a chromosome at random and then inverts the sub-string between these two positions. We illustrate the inversion mutation operation as follows:

Selected sub string

| | | | | | | | | |
|---|---|---|---|---|---|---|---|---|
| Parent | 4 | 6 | 3 | 5 | 7 | 1 | 8 | 2 |
| offspring | 4 | 6 | 7 | 5 | 3 | 1 | 8 | 2 |

**Figure 2.** Example of inversion mutation

### *3.3. Evaluation and Selection*

When using GA, we have to assets each chromosome on how well it fits with the problem requirements. Here, we use the makespan as the fitness value. The selection process is also known as an essential step in applying GA. The main objective is to guide in determining the chromosome for the next population. The selection process is done based on the fitness value. There have been many selection strategies reported for various GA applications [5]. We adopt the elitist selection strategy by selecting the best *pop_size* chromosome to the next generation.

## 4. Experimental Design and Results

There are two purposes of this section: First is to explain the design of the numerical experiments, including the design of the test problems and parameter setting. The second is to evaluate the effectiveness and efficiency of GAs to solve FSSP.

### *4.1. Design of Test Problems*

Several numerical experiments has been done to show the effectiveness and efficiency of the proposed approaches. In our experiments, we used a total of 18 different size Benchmark instances provided by literature [10]. Those problems have the number of jobs 20-100 and the number of machines 5-20. Those approaches were implemented in C++ and run on PC with processor Intel-Core i5. We set the crossover and mutation probabilities as 0.4 and 0.2, respectively.

The population size is varied based on the size of the problems. Table 1 summarizes the overall results of the experiments.

**Table 1:** Design of numerical experiment and results

| No. | Test problems | Number of jobs | Number of machines | max_gen | Optimal | mb-GA | jb-GA | NEH* | time** |
|---|---|---|---|---|---|---|---|---|---|
| 1 | ta001 | 20 | 5 | 600 | 1278 | 1376 | 1297 | 1286 | 25.18 |
| 2 | ta006 | 20 | 5 | 600 | 1195 | 1373 | 1195 | 1228 | 25.18 |
| 3 | ta011 | 20 | 10 | 600 | 1582 | 1716 | 1592 | 1680 | 33.88 |
| 4 | ta016 | 20 | 10 | 600 | 1397 | 1515 | 1412 | 1453 | 33.88 |
| 5 | ta021 | 20 | 20 | 600 | 2297 | 2346 | 2316 | 2410 | 36.83 |
| 6 | ta026 | 20 | 20 | 600 | 2226 | 2302 | 2230 | 2349 | 36.83 |
| 7 | ta031 | 50 | 5 | 600 | 2724 | 2899 | 2729 | 2733 | 34.36 |
| 8 | ta036 | 50 | 5 | 600 | 2829 | 3068 | 2832 | 2850 | 34.36 |
| 9 | ta041 | 50 | 10 | 700 | 3025 | 3459 | 3098 | 3146 | 40.91 |
| 10 | ta046 | 50 | 10 | 700 | 3006 | 3470 | 3116 | 3178 | 40.91 |
| 11 | ta051 | 50 | 20 | 700 | 3875 | 4192 | 3995 | 4038 | 49.56 |
| 12 | ta056 | 50 | 20 | 700 | 3698 | 4080 | 3829 | 3918 | 49.56 |
| 13 | ta061 | 100 | 5 | 700 | 5493 | 5646 | 5495 | 5567 | 47.82 |
| 14 | ta066 | 100 | 5 | 700 | 5135 | 5553 | 5144 | 5139 | 48.8 |
| 15 | ta071 | 100 | 10 | 800 | 5770 | 6366 | 5842 | 5848 | 63.23 |
| 16 | ta076 | 100 | 10 | 800 | 5303 | 5965 | 5344 | 5373 | 63.23 |
| 17 | ta081 | 100 | 20 | 2000 | 6286 | 6974 | 6456 | 6661 | 187 |
| 18 | ta086 | 100 | 20 | 2000 | 6437 | 7074 | 6661 | 6761 | 187 |

*Nawaz, Encore and Ham (NEH) [11]
**Computational Time of jb-GA (in the second)

The above results that jb-GA outperforms mb-GA on the solution quality, all of the time. It can reach the optimal/near-optimal solutions to the problem. The comparison with the results of the NEH algorithm shows an impressive performance of the jb-GA on the quality of solutions (95 percent). The above results also indicate the reasonable computational time of jb-GA.

In these experiments, we define the error as the percentage of (Obtained Solution – Optimal Solution)/Optimal Solution. The next Figure 3 illustrates the comparison of the errors for each instance.

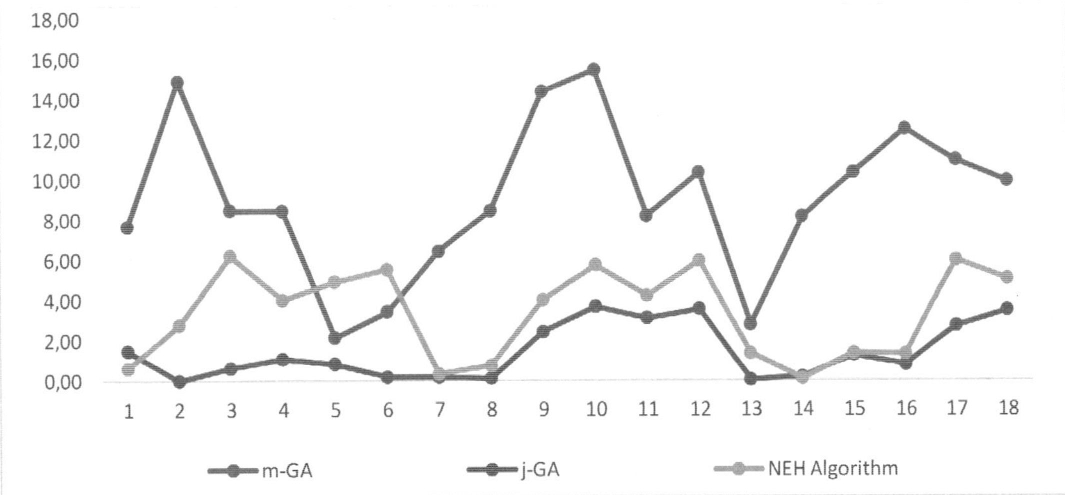


**Figure 3.** The comparative error of the methods

Like many other heuristic methods, the above results show that jb-GA still has a limitation on the number of instances solved (NIS) optimally. For some significant size problems, it even often reaches the near-optimal solutions. So there is always a place for improvement in the quality of solution or computational cost. Finally, in the next Figure 4, we illustrate the obtained schedule and the convergence of objective value in the generation for test problem ta061.

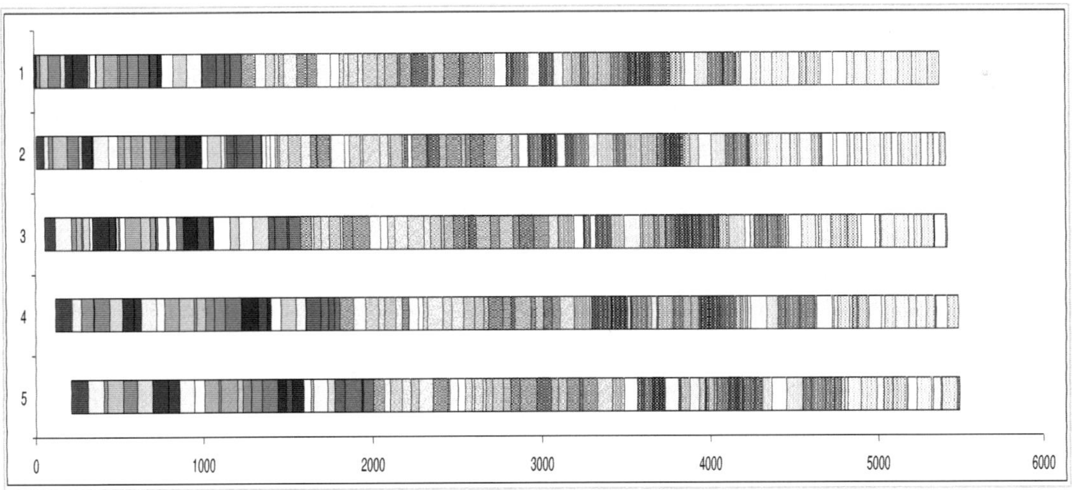


**Figure 4.** Makespan of the schedule for test problem ta061

## 5. Conclusions

We have presented two GA approaches called job-based GA (jb-GA) and machine-based GA (mb-GA) to solve FSSP. To demonstrate the effectiveness of the methods, we conducted several numerical experiments. We use the Benchmark scheduling test problems given in the literature. We compare the results with those provided by another heuristic algorithm (NEH algorithm). The results demonstrate that jb-GA has higher accuracy and achieves the optimal solution with reasonable computational time. This finding confirms the usefulness of GA to solve FSSP. The current limitation of this approach, like many other heuristic methods, is that jb-GA still sometimes gives the near-optimal solution.

**Acknowledgement**
This research has been supported by the Scientific Research Unggulan, Grant-in-Aid for Scientific Research by the Ministry of Research and Higher Education, Lampung University, No. : 2226/UN26.21/PN/2019, Indonesia, 2019.

**References**

[1] C. Rajendran and D. Chaudhuri, 1993 "An efficient heuristic approach to the scheduling of jobs in a Flow Shop. European," *J. Oper. Res.*, **61**, no. 3, pp. 318-325.,
[2] M. Basseur, F. Seynhaeve, and E. Talbi, 2002 "Design of multi-objective evolutionary algorithms: application to the flow-shop scheduling problem," in *Proc. of the 2002 Congress on Evolutionary Computation*, pp. 2 1151-1156.
[3] P. Jin and S. Kaoping, 2003 "Fuzzy flow-shop scheduling models based on credibility measure," in *Proc. of the 12th IEEE International Conference on Fuzzy Systems*, pp. 139-144,
[4] M. Gen and R. Cheng, 2000 *Genetic Algorithms and Engineering Optimization,* New York: John Wiley & Sons.
[5] M. Gen and R. Cheng, 1997 *Genetic Algorithms and Engineering Design.*
[6] A. Syarif, Y. S. Yun, and M. Gen, 2002 "Study on multi-stage logistic chain network: A spanning tree-based genetic algorithm approach," *Int. J. Comput. Ind. Eng.*, **43**, no. 1-2, pp. 299–314,
[7] A. Syarif and M. Gen, 2003 "Solving exclusionary side constrained transportation problem by using a hybrid spanning tree-based genetic algorithm," *J. Intell. Manuf.*, **14**, no. 3-4, pp. 389–399
[8] M. Gen and A. Syarif, 2005 "Hybrid genetic algorithm for multi-time period production/distribution planning," *Int. J. Comput. Ind. Eng.*, **48**, pp. 799-809,
[9] M. Gen and A. Syarif, 2003 "Double Spanning Tree-Based Genetic Algorithm for Two-Stage Transportation Problem," *Int. J. Knowledge-based, Eng. Syst.*, 7, no. 4, pp. 214-221
[10] E. Taillard, "Benchmarks for the basic scheduling problems., 1993 " *Eur. J. Oper. Res.*, **64**, pp. 278–285
[11] P. Chang, C. Liu, and C. Fan, 2006 "A Depth-First Mutation-Based Genetic Algorithm for Flow Shop Scheduling Problems," in *Proc. of International Conference on Hybrid Information Technology*, pp. 25–32.
[12] M. Seda, 2007 "Mathematical Models of Flow Shop and Job Shop Scheduling Problems," *Int. J. Appl. Math. Comput. Sci.*, **4**, no. 41, pp. 241–246.
[13] J. H. Holland, 1975 *Adaptation in Natural and Artificial Systems.*
[14] D. Goldberg, 1989 *Genetic Algorithm in Search, Optimization, and Machine Learning,* Reading, MA: Addison-Wesley.
[15] Z. Michalewicz, 1989 *Genetic Algorithms + Data Structure = Evolution Program.* New York: Springer-Verlag.

# Adsorption isotherm of multicomponent solution of Cu(II) ions, crystal violet, and methylene blue on Silica-Magnetite *Porphyridium* sp. algae hybrid

**D Permatasari[1], Buhani[1,*], M Rilyanti[1], and Suharso[1]**

[1] Department of Chemistry, Faculty of Mathematic and Natural Sciences, University of Lampung, Prof. Soemantri Brojonegoro street number.1, Bandar Lampung,35145, Indonesia

* Corresponding Email: buhani@fmipa.unila.ac.id

**Abstract.** Algae-silica hybrid (HA-S) from *Porphyridium* sp. has been modified by coating particles of magnetite ($Fe_3O_4$) via a sol-gel process using (TEOS) to produce silica-magnetite algae hybrid (HA-SM). The produced material HA-SM was applied as adsorbent for the multicomponent solution adsorption of Cu(II) ions, crystal violet (CV), and methylene blue (MB). Identification of functional groups was performed by infrared spectrometer (IR) indicating that in HA-SM there are an active organic groups derived from *Porphyridium* sp., while silanol and siloxane derived from silica matrix. In addition, the results of the analysis using Scanning Electron Microscopy with Energy Dispersive X-Ray (SEM-EDX) showed HA-SM has more heterogeneous surface morphology. The adsorption model of the multicomponent solutions of Cu(II) ions, CV, and MB by HA-S and HA-SM tend to follow the model of Freundlich adsorption isotherm.

**Keywords:** *Porphyridium sp.*, HA-SM, adsorption, Freundlich

## 1. Introduction

Water pollution caused by heavy metals and dyes tends to increase [1-3], especially those originating from wastewater resulting from industrial activities that have an impact on environmental pollution with amount of more than $7 \times 10^5$ tons heavy metals could not be degraded in environment [4]. In addition, it could be both harmful to a variety of living species and easily to be accumulated through the food chain which may eventually affect human beings, more stable and dangerous due to the formation of more complex chemical structure and difficult to degrade [5].

The most widely used dyestuffs are crystal violet (CV) and methylene blue (MB) due to their properties with good solubility. The CV and MB are a basic aniline dyes with molecular formula $C_{25}H_{30}N_3Cl$ and $C_{16}H_{18}C_1N_3S$. The CV and MB dyestuffs are reported to cause a bad effects on humans, animals, and microorganisms as well as increase the environmental pollution. In addition, the CV and MB dyestuffs are an aromatic compound with a complex structure and hard to degrade. One of the heavy metal resulted from industrial waswater is Cu metal. Cu metal is needed by human beings for Fe metabolism in hemoglobin, but in high concentration, it will be toxic and accumulated in the human body [6,7].

Therefore, it is important to control the discharge of CV and MB dyestuffs also Cu(II) ions into the wastewater before they are being released into the environment effectively with process the wastes containing the dyestuffs and heavy metals before their discharged into the environment. Dyes waste water treatment can be done using adsorption method because it is simple, relatively inexpensive, and does not generate byproducts that are harmful to the environment [8,9]. The success of the adsorption process is determined by the compatibility of the nature and type of the adsorbent used. The effective adsorbent for adsorption is the adsorbent which has a large rate and adsorption capacity, chemical stability, environmental friendly, and it can be used repeatedly. In general, filtration and centrifugation are used after adsorption process. They are time consuming. The blockage of filters and the loss of adsorbents may occur in filtration method. The use of magnetic separation method offers an alternative way to solve these problems [10].

Several studies have been reported on the preparation of adsorbent and their adsorption of dyestuffs solutions, which consisted material of $Fe_3O_4$ [1], $\gamma$-$Fe_2O_3$ [2], and nanocomposite mesoporous silica. In order to increase the adsorption capacity of the magnetic adsorbent, the magnetic adsorbent synthesis can be combined by adding an active group derived from algae biomass. Algae biomass is naturally an effective adsorbent for absorbing pollutants in both inorganic compounds such as heavy metals [11] and organic compounds derived from dyestuffs [4]. Several studies have shown that algae are alive and in the form of their biomass can adsorb chemical compounds, especially those containing cations such as metal ions and organic cations. The adsorption ability of algae biomass as an adsorbent can be increased by immobilizing algae biomass using various supporting polymers which are generally inorganic solids such as silica [12]. Enhancement of adsorption rate from adsorbate on the adsorbent was carried out using magnetite particle coating technique on silica as a biomass support matrix [13].

In this study, HA-SM have been synthesized through a sol-gel process using tetraethyl orthosilicate (TEOS) as the precursor and organic group of *Porphyridium* sp. algae biomass and coated with magnetite particle. The synthesis of this adsorbent has never been done by previous researchers. The study of the adsorption of HA-SM material on the multicomponent solutions of Cu(II) ions, CV, and MB discuss about the adsorption isotherm of the absorbent to the multicomponent solutions of Cu(II) ions, CV, and MB. The HA-SM material is effective as an adsorbent for the removal of the multicomponent solutions of Cu(II) ions, CV, and MB, especially for the water treatment process from dyestuffs and heavy metals contaminants.

## 2. Experimental methods

### 2.1. Materials

*Porphyridium* sp. biomass was obtained from the Lampung Sea Cultivation Bureau (Balai Besar Pengembangan Budidaya Laut Lampung, Indonesia). All chemicals used in the research were purchased from Merck Co., Inc. (Germany) and consisted of TEOS; distilled water; HCl 0.1 M; $NH_4OH$ 1 M; NaOH 0.1 M; $FeSO_4 \cdot 4H_2O$; $FeCl_3 \cdot 6H_2O$; MB; CV; $CuSO_4 \cdot 5H_2O$; ethanol; $KNO_3$ 1 M; $CH_4N_2O$ 0.1 M; and $Na_2EDTA$ 0.1 M. $Fe_3O_4$ particel used in this study were synthesized based on procedures performed by previous researchers [13].

### 2.2 Preparation of HA-SM adsorbent

The HA-SM synthesis was carried out by interacting 5 mL of TEOS, 2.5 mL of distilled water, and 0.1 g of magnetite particle in a plastic bottle by stirring for 30 min. During stirring, 1M solution of HCl up to pH $\approx$ 2 (solution A). In a different plastic bottle, 0.1 g *Porphyridium* sp. biomass was mixed with 5 mL of ethanol, and the mixture was stirred for 30 min (solution B). By stirring, solution A was mixed with solution B to obtain a gel. The gel produced was left for 24 h, cleaned by water and ethanol up to the pH of filtrate was near 7, and then dried in an oven at 40 ºC until its weight was constant. The dried material was ground and sieved through a sieve of 200 mesh. Synthesis of HA-S was carried out by a similar procedure applied to yield HA-SM, but without adding the magnetite particle [14].

### *2.3 Characterization of HA-S and HA-SM adsorbent*

The synthesis resulted in a material HA-S and HA-SM that was characterized by an infrared (IR) spectrometer Prestige-21 Shimadzu, Japan to identify functional groups that existed in the material. The surface morphology and constituent elements were investigated by Scanning Electron Microscopy with Energy Dispersive X-Ray (SEM-EDX) (Zeiss MA10, Germany).

### *2.4. Batch adsorption experiment*

Batch adsorption experiments were carried out to estimate the efficiency of HA-S and HA-SM for the removal of the multicomponent solutions of Cu(II) ions, CV, and MB from the artificially contaminated water. Stock solution of the multicomponent solutions of Cu(II) ions, CV, and MB of 1; 2.45; and 3.13 mmol $L^{-1}$ respectively was prepared by dissolving accurately 1 g in 1 L of distilled water from which the solutions of desired concentrations were prepared by dilution. The adsorption process by batch method was carried out by interacting 25 mL of the multicomponent solutions of Cu(II) ions, CV, and MB with pH = 8, and then adsorbent stirred at 15 min contact time and 100 rpm speed to determine the influence of solution concentration variation.

Concentration of multicomponent solution of Cu(II) ions, CV, and MB before and after adsorption processes is specified using AAS (Atomic Absorption Spectrophotometer) and UV-Vis Spectrophotometer (Agilent Cary 100, U.S.A.) at a wavelength of 591 (CV) and 664 nm (MB). The concentration of the multicomponent solutions of Cu(II) ions, CV, and MB adsorbed per unit mass of adsorbent and the percentage of the multicomponent solutions of Cu(II) ions, CV, and MB adsorbed was determined using Eqs. (1)-(2).

$$qe = \frac{(Co - Ce)}{m} \times V \quad (1)$$

$$qt = \frac{(Co - Ct)}{m} \times V \quad (2)$$

Where $C_o$ is concentration in the initial state, $C_e$ is concentration at equilibrium, $C_t$ is concentration at a certain time t of solution (mg $L^{-1}$), m is the mass of adsorbent (g), V is the volume of the solution (L), q is the number of solution adsorbed per unit mass (mg $g^{-1}$).

### *2.5. Adsorption mechanism*

Determination of adsorption mechanism was done using sequential desorption method, on the multicomponent solutions of Cu(II) ions, CV, and MB. It was started with the adsorption process using 0.1 g HA-S and HA-SM adsorbent and added with 25 mL the multicomponent solutions of Cu(II) ions, CV, and MB in the optimum concentration with the contact time of 15 min at 100 rpm stirring speed. The resulted material from centrifugation process was washed by distilled water, $KNO_3$, $CH_4N_2O$, and $Na_2EDTA$ solution of 25 mL respectively.

## 3. Results and discussion

### *3.1. Characterization of adsorbent*

Synthesis of HA-S and HA-SM materials was performed through a sol-gel process using TEOS as a precursor and continued with coating by magnetite particle. Characterization of the materials of HA-S and HA-SM included identification of functional group, crystalline level, particle size distribution, surface morphology, and material composition. Identification of functional groups with IR spectroscopy has been performed to find out HA-S modifications that produce HA-SM containing magnetite particles (Figure 1). In Figure 1(a) it can be observed that the IR spectra of the

*Porphyridium* sp. algae biomass showed the presence of –OH group bound to C=O, amplified by the appearance of uptake in wavenumber 1651.07 $cm^{-1}$ which is a typical uptake of C=O of a carboxylic acid. In the wavenumber 3749.62 $cm^{-1}$ there is a primary N–H group. Uptake in wavenumber 2931.86 $cm^{-1}$ indicates the presence of a C–H group of a carbon (aliphatic) chain $CH_2$ [12].

In Figure 1(b) which is the IR spectra of HA-S and Figure 1(c) of HA-SM gives the same relatively IR uptake value that is the presence of strong uptake peaks in wavenumber 3425.58 $cm^{-1}$ indicated the presence of O-H vibration. Uptakes in the 1651.07 $cm^{-1}$ area show the O-H vibration of bound water molecules while 1049.28 $cm^{-1}$ is an asymmetry vibration of the siloxane group (Si–O–Si). The existence of C–H stretch vibration indicates the presence of an aliphatic group (–$CH_2$) in the wavenumber region 2931.8 $cm^{-1}$. The uptake at 856.39 $cm^{-1}$ explains the presence of symmetrical stretch vibration of siloxane groups from Si–O and in absorption regions 432.05 $cm^{-1}$ indicates the vibration of Si–O [21]. In IR spectra for HA-SM there is also a magnetite group indicated by wavenumber 570.93 $cm^{-1}$ overlapping with Si-O-Si uptake band in wavenumber 432.05 $cm^{-1}$ as well as there exists wavenumber 856.39 $cm^{-1}$ derived from Si–O stretch vibration [12].

SEM images and energy dispersive X-ray spectroscopy (EDX) spectra from HA-S and HA-SM are shown in Figure 2. The results of surface morphology analysis with SEM show morphological differences on surface HA-SM compared with HA-S, due to the presence of the magnetite particles. The HA-SM looks surface morphology which is more heterogeneous and contrast, this occurs because of the presence of Fe element derived from magnetite particles, while in the HA-S more dominated silica matrix so it is more homogeneous [15]. While the data EDX (Table 1) can also be known that on the HA-SM, besides there are constituent elements derived from the silica matrix and *Porphyridium* sp. biomass, there are also other elements that come from magnetite (Fe and O) as shown in Figure 2(b) and Table 1.

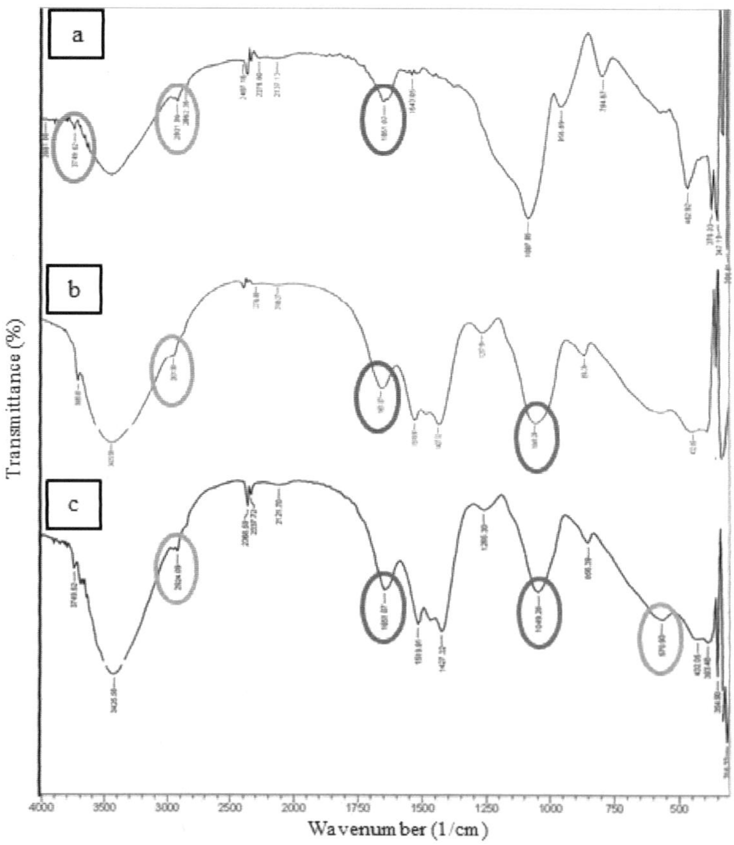


**Figure 1.** Infrared spectra of the (a) *Porphyridium* sp. (b) HA-S, and (c) HA-SM.

**Figure 2.** SEM images and EDX spectra of (a) HA-S and (b) HA-SM.

*3.2. Adsorption*

*3.2.1. The influence of multicomponent solution of Cu(II) ions, CV, and MB concentration*

The HA-S and HA-SM adsorption capability in absorbing the multicomponent solution Cu(II) ions, CV, and MB were investigated through the interaction of solutions at varying concentrations ranging from 0.1– 0.5 mmol $L^{-1}$. In Figure 3, it can be observed that there is an increase in the number of multicomponent solution Cu(II) ions, CV, and MB adsorbed with increasing concentration of solution. Adsorption of multicomponent solution of Cu(II) ions, CV, and MB increased sharply at low equilibrium concentrations and gradually increased at high concentrations. This shows that the HASM have a high adsorption affinity. In order to obtain the adsorption isotherm pattern that plays a role in describing the distribution between solution and solid phase, the data contained in Figure 3 were analyzed using adsorption isotherm models of Langmuir [Eq. (3)] and Freundlich [Eq. (4)] [8,14] as follows:

$$\frac{1}{q_e} = \frac{1}{q_m K_L C_e} + \frac{1}{q_m} \quad (3)$$

$$\log q_e = \log K_F + \frac{1}{n \log C_e} \quad (4)$$

Where $C_e$ (mg $L^{-1}$) is the equilibrium concentration of solution, $q_e$ (mg $g^{-1}$) is the adsorption capacity of solution at equilibrium, $q_m$ (mg $g^{-1}$) is the adsorbent monolayer adsorption capacity, and $K_L$ is the equilibrium constant including the affinity of binding sites (L $mg^{-1}$). $K_L$ and $q_m$ can be determined from the linear equation plot of log $1/q_e$ vs $1/C_e$ which will produce a straight line with $1/q_m K_L$ as slope and $1/q_m$ as intercept. $K_F$ ((mg $g^{-1}$) (L $mg^{-1})^{1/n}$) is the adsorption capacity factor and $n$

is the adsorption intensity factor, with $n$ values ranging from 1 to 10 [14]. Next, plot of log $q_e$ vs log $C_e$ will produce $K_F$ and exponent of $n$. Langmuir adsorption isotherm model assumes that on the surface of the adsorbent, there are a certain number of active sites that are proportional to the surface area, the adsorbent surface is uniform, and the adsorption process is monolayer. While Freundlich adsorption isotherm model is an empirical equation employed for heterogeneous system and adsorption at multilayers [15,16]. The adsorption parameters of multicomponent solution Cu(II) ions, CV,and MB by HA-S and HA-SM shown in Table 1. Show that the $R^2$ value of the Freundlich adsorption isotherm model is higher than that of Langmuir, adsorption isotherm model for adsorption of multicomponent solution Cu(II) ions, CV, and MB both in HA-S and HA-SM [11,12].

This suggests that the adsorption model for multicomponent solution of Cu(II) ions, CV, and MB are more fitting to the Freundlich adsorption isotherm model supported by the adsorption of the experimental results closer to the Freundlich model (Figure 3). Similar results were reported for the adsorption isotherm of various organic pollutants onto activated carbon cloth. Freundlich constants ($n$) of multicomponent solution of Cu(II) ions, CV, and MB by HA-S and HA-SM (Table. 1) have values greater than 1. This indicates that the adsorption of multicomponent solution Cu(II) ions, CV, and MB on HA-S and HA-SM is favorable in this research. The multicomponent solution of Cu(II) ions, CV, and MB adsorption in HA-S and HA-SM not only occurs on active sites on the homogeneous (monolayer) adsorbent surface but also occurs reversibly on heterogeneous (multilayer) adsorbent surfaces. The presence of magnetite particle in HA-S causes adsorbent to be magnetite by increasing the removal rate of multicomponent solution of Cu(II) ions, CV, and MB from solution and the amount of multicomponent solution Cu(II) ions, CV, and MB adsorbed.

In addition, the presence of magnetite particle causes the particle size of HA-SM to be smaller than that of the HA-S. This increases the surface area on HA-SM so that the number of active sites will also increase [17]. Thus, it can be stated that the adsorption of multicomponent solution of Cu(II) ions, CV, and MB on HA-S and HA-SM not only occurs chemically through the interaction of anionic active groups with cationic multicomponent solution Cu(II) ions, CV, and MB but also occurs through the physical interaction due to the influence of the magnetite properties of the adsorbent.

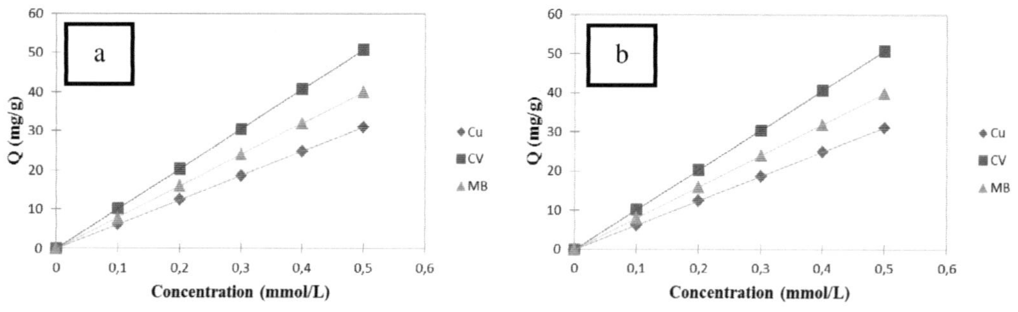


**Figure 3.** Influences of solutions concentration on adsorption multicomponent solution of Cu(II) ions, CV, and MB by (a) HA-S and (b) HA-SM (pH 8 and contact time 15 min).

**Table 1.** Langmuir and Freundlich parameters for the adsorption of multicomponent solution Cu(II) ions, CV, and MB on HA-S and HA-SM.

| Adsorbent | | Isotherm Model | | | | | |
|---|---|---|---|---|---|---|---|
| | | Langmuir | | | Freundlich | | |
| | | $n_m$ (mg/g) | $K_L$ (L/mg) | $R^2$ | n | $K_F$ (mg/g) $(L/mg)^{1/n}$) | $R^2$ |
| HA-S | Cu(II) | 32.894 | 0.573 | 0.756 | 0.672 | 38.832 | 0.982 |
| | CV | 105.263 | 0.730 | 0.492 | 1.397 | 44.565 | 0.911 |
| | MB | 243.902 | 0.317 | 0.006 | 0.884 | 108.292 | 0.651 |
| HA-SM | Cu(II) | 227.272 | 0.366 | 0.026 | 0.970 | 96.360 | 0.907 |
| | CV | 121.951 | 0.650 | 0.972 | 1.267 | 49.808 | 0.993 |
| | MB | 52.356 | 6.586 | 0.530 | 1.655 | 79.341 | 0.859 |

*3.2.2. Types of interactions*

The occurred adsorption mechanisms show type of interactions in HA-S and HA-SM adsorbents to multicomponent solutions of Cu(II), CV, and MB was dominated by complex formation interactions (Figure 4.). The desorption percentage of $Na_2EDTA$ eluent on HA-S adsorbents is 69.29; 85.63; and 67.43%, while on HA-SM is 62.89; 52.41; and 71.71% for Cu(II), CV, and MB respectively. The complex formation interaction between HA-S and HA-SM with the multicomponent solutions of Cu(II) ions, CV, and MB is caused by multicomponent solutions formed on the HA-S and HA-SM adsorbent surfaces due to the ionic process [12].

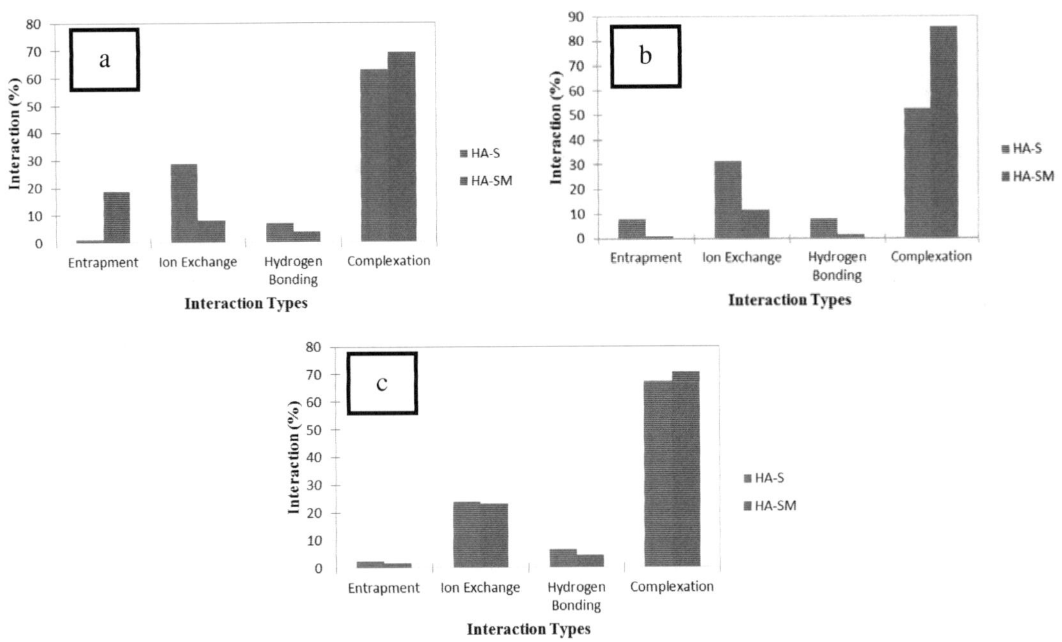


**Figure 4.** Types interaction HA-S and HA-SM the multicomponent solutions of (a) Cu(II) ions, (b) CV, and (c) MB.

## 4. Conclusions

Modification of HA-S derived from *Porphyridium* sp. algae biomass with the coating magnetite particle has produced an effective adsorbent to remove Cu(II) ions, CV, and MB in solution. Characterization of HA-SM material shows some active groups acting as active sites for binding of Cu(II) ions, CV, and MB. Coating magnetite particle increases the rate and capacity of the HA-S adsorption to the multicomponent solutions of Cu(II) ions, CV, and MB. In the HA-SM Materials, it is also proven that there are more complex bond interactions. Thus, HA-SM is an effective adsorbent for removing Cu(II) ions, CV, and MB in solution, so it can be applied to the processing of wastewater containing dyestuffs and heavy metals.

## Acknowledgement

The authors would like to thank the Directorate of Research and Community Service, Directorate General for Research and Development, Ministry of Research, Technology and Higher Education of the Republic of Indonesia who have funded this research in accordance with contract number 065/SP2H/LT/DPRM/2019. The authors give a high appreciation to Technical Service Unit of the Integrated Laboratory and The Technology Innovation Center-University of Lampung (UPT Laboratorium Terpadu dan Sentra Inovasi Teknologi Universitas Lampung).

## References

[1] Shao Y, Zhou L, Bao C, Ma J and Wang F 2016 *J. Chem. Eng.* **283** 1127-1136
[2] El-Qanni A, Nassar N N, Vitale G and Hassan A 2016 *J. Colloid Interface Sci.* **461** 396-408
[3] Fu J, Xin Q, Wu X, Chen Z, Yan Y, Liu S, Wang M and Xu Q 2016 *J. Colloid Interface Sci.* **461** 292-304
[4] Daneshvar E, Vazirzadeh A, Niazi A, Kousha M, Naushad M and Bhatnagar A 2017 *J. Clean. Prod.* **152** 443-453
[5] Martins R J E, Pardo R and Boaventura R A R 2004 *Water. Res.* **38** 693-699
[6] Celeki A and Bozkurt H 2013 *Environ. Sci. Pollut. Res.* **20** 7355-7366
[7] Dardouri S and Sghaier J 2017 *Chin. J. Chem. Eng.* **25** 1282-1287
[8] Buhani, Narsito, Nuryono, Kunarti E S and Suharso 2015 *Desalin. Water. Treat.* **55** 1240-1252
[9] Suharso and Buhani 2011 *Asian. J. Chem.* **23** 1112-1116
[10] Guler U A, Ersan M, Tuncel E and Dugenci F 2016 *Saf. Environ. Prot.* **99** 194–206
[11] Buhani, Musrifatun, Pratama D S, Suharso and Rinawati 2017 *Asian. J. Chem.* **29(12)** 2734-2738
[12] Schneider, J F, Saeki M J, PedrosaV A, Yoshito W K, Colauto F, Ortiz W A and Castro G R, 2016 *Appl. Surf. Sci.* **367** 533–541
[13] Buhani, Rinawati, Suharso, Yuliasari D P and Yuwono S D 2017 *Desalin. Water. Treat.* **80,** 203-213
[14] Buhani, Hariyanti F, Suharso, Rinawati and Sumadi 2019 *Desalin. Water. Treat.* ***142*** *331-340*
[15] Buhani, Suharso and Satria H 2011 *Eur. J. Sci. Res.* **51(4)** 467-476
[16] Buhani, Suharso and Fitriyani A Y 2013 *Asian. J. Chem.* **25(5)** 2875-2880
[17] Buhani, Herasari D, Suharso and Yuwono S D 2017 *Orient. J. Chem.* **33(1)** 418-429

# Flower baggings in affecting mangosteen fruit qualities at harvest and during storage

**S E Widodo[1*], M Kamal[1], Zulferiyenni[2], D Chandra[3], and D W Kusuma[3]**

[1] Department of Agronomy and Horticulture, Faculty of Agriculture, University of Lampung, Sumantri Brojonegoro Street No.1, Bandar Lampung 35145, Indonesia
[2] Department of Agricultural Product Technology, Faculty of Agriculture, University of Lampung, Sumantri Brojonegoro Street No.1, Bandar Lampung 35145, Indonesia
[3] Department of Post-graduate Studies of Agronomy, Faculty of Agriculture, University of Lampung, Sumantri Brojonegoro Street No.1, Bandar Lampung 35145, Indonesia

*Email: sestiwidodo@gmail.com

**Abstract.** Physiological causes and insect attact are believed to increase yellow latex exudates in mangosteen fruits. To inhibit the causes, flower bagging should be applied. This research was aimed at studying the effects of flower baggings to two different flower developments in affecting mangosteen fruit qualities at harvest and during storage. Three bagging materials (unbagged, paper, and baloon) were applied to flowers of 2 and 4 weeks after anthesis (WAA). The fruits were sampled every 2 weeks during the periods of 8-16 WAA. The results showed that except α-mangosteen content that was slightly decreased during the latest periode of fruit growth by bagging at preharvest, flower baggings of both bagging materials and application periods mostly did not affect mangosteen fruit qualities at harvest, but they affected fruit shelf-life and qualities during storage. Flower baggings resulted in increased fruit shelf-life, with paper bagging applied in 2 WAA was better than that applied in 4 WAA. Paper bagging in 2 WAA resulted in the mangosteen fruit shelf-life of 29 days compared to 4 WAA which resulted in 14 days shelf-life. This research proved also that the occurence of yellow latex was much more likely affected by physiological causes, not by insect attacts.

**Keywords**: bagging, fruit quality, harvest, mangosteen, storage

## 1. Introduction

Among the 20 known species in the genus *Garcinia* [1], mangosteen (*Garcinia mangostana* L.) is the most studied fruit. That is because of its very wide use, from consumption as medicine to table fruit which is consumed fresh or minimally processed [2, 3]. Known as the "Queen of Tropical Fruits", mangosteen is classified as a high-value fruit crop. Increasing the quantity and quality of the fruit must be maintained since it is still in the tree (pre-harvest) until treatment at harvest and post-harvest [4].

Unfortunately, the quality of mangosteen fruit is also known to often experience postharvest damage, due to insect attack, physiological damage, and poor postharvest handling [5]. Physiological causes, insect attact, and improper fruit handling are commonly believed to increase the occurence of yellow latex (gamboge) disorder in mangosteen fruits. Gamboge or yellow latex disorder is the type of postharvest damage or physiological disease that is considered to be the most detrimental. This is mostly true if the yellow sap has contaminated the fruit flesh, so it will taste bitter. This type of damage is quite difficult to detect, because although the causes are studied, the method of detection is still unknown. So far, the yellow latex disorder is known to be due to two contributing factors, which are related to water content and Ca deficiency [6–11]. Visual detection is still difficult, because the yellow sap that is seen on the surface of the rind pericarp often does not prove that the fruit flesh has also been contaminated with this bitter yellow gum. Non-destructive detection efforts have been widely tested [12–14], unfortunately the results are generally still felt to be ineffective and economically unbeneficial.

To inhibit physiological causes and insect attact that led to yellow latex disorder, flower bagging should be applied. It is usually done after the flowers are completely open (anthesis) [15]. This is useful in the efficiency of the bagging material which will be wasted if the flowers fall before they develop. The bagging material also affects the physical properties of the fruit, bywhich bagging with cement paper was reported to be the best [16, 17, 18]. It is known that fruit bagging can affect the intensity of pest attacks, fruit quality, physical and chemical properties of fruit [4, 19, 20], but information about the bagging treatment of mangosteen is difficult to obtain. Therefore, this research was aimed at studying the effects of flower baggings to two different flower developments in affecting mangosteen fruit qualities at harvest and during storage.

## 2. Materials and Methods

This research that was conducted in July–December 2017, consisted of two consecutive research, namely field research and laboratory postharvest research.The field research was conducted in a farmer's field at Gisting village, Tanggamus district, Lampung Province, Indonesia. The mangosteen crop samples were about 38 years old, and located at -5°27'30" NL 104°42'8" SL, ± 537.1 m above-sea-level. Fruit samples were analyzed in (1) the Laboratory of Horticultural Postharvest, (2) the Laboratory of Plant Insects and Diseases, and (3) Biotechnology Laboratory, Fac. of Agriculture, University of Lampung, Bandar Lampung, Indonesia, (4) the Laboratory of Pharmacy Analysis and Medicinal Chemistry, Fac. of Pharmacy, University of Pajajaran, Bandung, Indonesia, and (5) the Integrated Laboratory and Center for Technology Innovation, University of Lampung, Indonesia. The research was started by tagging mangosteen flower at anthesis.

The field research used Completely Randomized Design (CRD) with three replications in each sampling periode of five samplings totally. It was arranged in a 2 × 3 factorial design. The first factor was bagging date [2 and 4 weeks after anthesis(WAA)], and the second one was bagging material (unbagged or control, banana 'Cavendish'- paper bag, and baloon). The reused banana 'Cavendish'-paper bags were received from Great Giant Foods, Co. Ltd., Terbanggi Besar, Central Lampung through Nusantara Tropical Farm, Co. Ltd., Labuhan Ratu, East Lampung, Indonesia. Three bagging materials (unbagged, banana 'Cavendish'- paper bag, and baloon) were applied to flowers of 2 and 4 WAA. The fruits were then sampled every 2 weeks during the fruit development periods of 8-16 WAA. Observations to fruit variables (fruit diameter, weight, temperature, and α-mangosteen content in the rind, and yellow latex spots on the surface rind pericarp) were conducted in every two weeks sampling in the sampling periods of 8-16 WAA. Fruit surface temperature was taken with an infrared thermometer. The α-mangosteen content was analyzed with HPLC [Dionex-UltiMate® 3000, autosampler, column compartment, Ultimate 3000 pump, UV detector, column Enduro C-18 (250 mm × 4.6 mm, 5 μm) with C18 guard] based on [21]. At the end of sampling period of 16 WAA, the fruit variables of °Brix, free acid content, and sweetness level were analyzed. In addition, the data of rainfall, and insects trapped on the yellow-sticky insect trapper were also taken. The insects trapped were then identified based on [22] in the Laboratory of Plant Insects and Diseases, Fac. of Agriculture, University of Lampung, Bandar Lampung, Indonesia.

The laboratory postharvest research used the same 2 × 3 factorial design as in the field research with three replications. Fruits from the last sampling (121 days; stage 0 yellowish white or yellowish white with light green)[23] were then brought to the Laboratory of Horticultural Postharvest, Fac. of Agriculture, University of Lampung, Bandar Lampung, Indonesia. The samples were put in a storage room of 27-28 °C, and the observations were terminated when the fruits reached stage 6 (purple black color) [23]. The observed variables were shelf-life, fruit weight, fruit rind color, dissolved solid content (°Brix), free acid content, sweetness level, yellow latex spot, rind weight, rind thickness, fruit diameter, aryl weight, microscopic transverse observation of mangosteen rind. The microscopic transverse observation was analyzed based on [7] in Biotechnology Laboratory, Fac. of Agriculture, and the Integrated Laboratory and Center for Technology Innovation, University of Lampung, Indonesia.

Data were analyzed statistically with an orthogonal polynomial contrast at 5% level (SAS System for Windows V9.1), and then presented into tables and line graphs.

## 3. Results and Discussion

The results showed that flower baggings mostly did not affect mangosteen fruit development. The fruit grew quickly up to 12 WAA and then slowed down to 16 WAA (figure 1-A), regardless of baggings. This same phenomenon of fruit weight increase (figure 1-A) was noted also with fruit diameter (data were not shown). In general, therefore, flower baggings of both bagging materials and application periods mostly did not affect mangosteen fruit qualities at harvest, such as °Brix, free acid content, and sweetness level (table 1). No differences in fruit surface temperature (figure 1-B) and wet season during fruit sampling (10-20 rainy-days), escpecially in November 2017 that was classified as higher than normal (301-400 mm), might support fruit growth, regardless of bagging materials and application periods. Similar results of bagging were reported with other fruit [24].

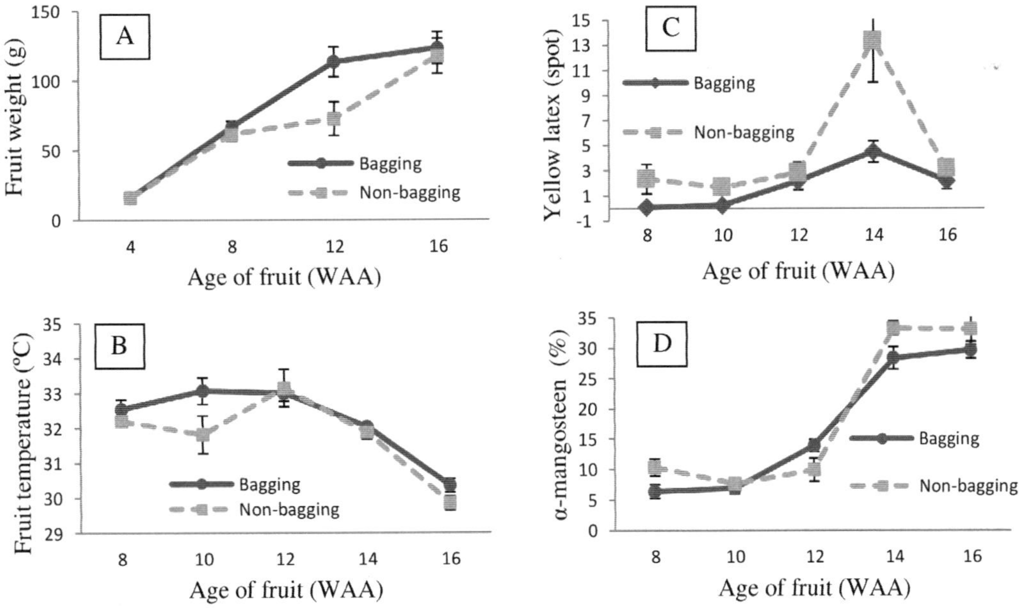


**Figure 1.** Effect of bagging on fruit weight, yellow latex, fruit temperature, and α-mangosteen of mangosteen fruit.

The yellow latex spots were present throughout fruit growth, irrespective of bagging applications (figure 1-C), even though the bagging fruits experienced lower incidence of yellow latex spots, and fruits bagged with balloon experienced the worst incidence of yellow latex spots. By considering the insects trapped on the yellow-sticky insect trapper, which were dominated by black ants (1.960 *Dolichoderus thoracicus*), compared with *Bactrocera dorsalis* (23) and *Nilaparvata lugens* (4), yellow latex disorder was much more likely affected by physiological causes, not by insect attacks [25].

**Table 1.** Effects of bagging materials and dates on soluble solid content (°Brix), acid content, and sweetness of mangosteen fruit at harvest based on orthogonal contrast[a].

| Contrast | °Brix (%) | Acid content (g/100 g) | Sweetness |
|---|---|---|---|
| **Control vs Bagging** | 14.03 vs 13.53 (0.6282) | 0.14 vs 0.15 (0.7951) | 169.22 vs 135.10 (0.4715) |
| **Baloon vs Paper** | 13.90 vs 13.16 (0.5397) | 0.14 vs 0.15 (0.8807) | 113.36 vs 156.86 (0.4280) |
| **Baggings 2 vs 4 WAA** | 13.50 vs 13.56 (0.9552) | 0.11 vs 0.19 (0.1928) | 160.76 vs 109.43 (0.3520) |
| **Baloons 2 vs 4 WAA** | 14.46 vs 13.33 (0.5034) | 0.17 vs 0.11 (0.4751) | 87.92 vs 138.79 (0.5103) |
| **Papers 2 vs 4 WAA** | 12.53 vs 13.80 (0.4556) | 0.05 vs 0.26 (0.0197) | 233.60 vs 80.07 (0.0631) |

[a]Values inside parentheses are P-contrast values; bagging = banana 'Cavendish'- paper bag and baloon; paper = banana 'Cavendish'- paper bag; WAA = week after anthesis; Sweetness = °Brix:acid content ratio.

α-Mangosteen content was increased tremendously during 10-14 WAA, and again, regardless of bagging (figure 1-D), bagging materials and application periods (data were not shown). However, bagging slightly decreased α-mangosteen content during the latest periode of fruit growth (figure 1-D), regardless of bagging materials and application periods. Similar results were reported by other researchers [20, 26] and its increase was simply in paralel with anthosianin development [23].

Data in table 2 showed that mangosteen fruits that were bagged at preharvest had longer shelf-life by 6 days than control, and preharvest bagging at 2 WAA produced fruits that had longer shelf-life by 6 days than 4 WAA. Bagging with baloon was better than with banana 'Cavendish'- paper bag by 7 days storage, regardless of bagging date. However, when bagging was applied with banana 'Cavendish'- paper bag, bagging at 2 WAA produced fruits that had longer shelf-life by 15 days storage than 4 WAA.

**Table 2.** Effects of bagging materials and dates on shelf-life, weight loss, and sweetness of mangosteen fruit at storage based on orthogonal contrast[a].

| Contrast | Shelf-life (days) | Fruit weight loss (%) | Sweetness |
|---|---|---|---|
| **Control vs Bagging** | 18.5 vs 24.37 (0.0863) | 7.42 vs 12.26 (0.1431) | 189.30 vs 154.92 (0.4947) |
| **Baloon vs Paper** | 27.75 vs 21.00 (0.0876) | 18.62 vs 5.90 (0.0086) | 120.61 vs 189.23 (0.2557) |
| **Baggings 2 vs 4 WAA** | 27.50 vs 21.25 (0.1079) | 12.15 vs 12.36 (0.9516) | 200.33 vs 109.52 (0.1474) |
| **Baloons 2 vs 4 WAA** | 26.50 vs 29.00 (0.6125) | 19.91 vs 17.32 (0.6014) | 154.01 vs 87.22 (0.4203) |
| **Papers 2 vs 4 WAA** | 28.50 vs 13.50 (0.0185) | 4.39 vs 7.40 (0.5454) | 246.64 vs 131.82 (0.1877) |

[a]Values inside parentheses are P-*contrast values;* bagging = banana 'Cavendish'- paper bag and baloon; paper = banana 'Cavendish'- paper bag; WAA = week after anthesis; Sweetness = °Brix:acid content ratio.

A transverse observation of mangosteen mesocarp with SEM revealed that mesocarp cells bagged with baloon (C) were smaller and more compact that those unbagged (A) and bagged with banana 'Cavendish'- paper bag (figure 2). These smaller and more compact mesocarp cells of the fruits bagged with baloon might inhibit traspiration and resulted in longer shelf-life. Unfortunately, this longer shelf-life resulted in increasing fruit weight loss (table 1).

Fruits of all treatments during postharvest storage developed yellow latex spots on the rind surface. Just like in preharvest applications, this postharvest results suggested that yellow latex disorder was much more likely affected by physiological causes, not by insect attacts [25].

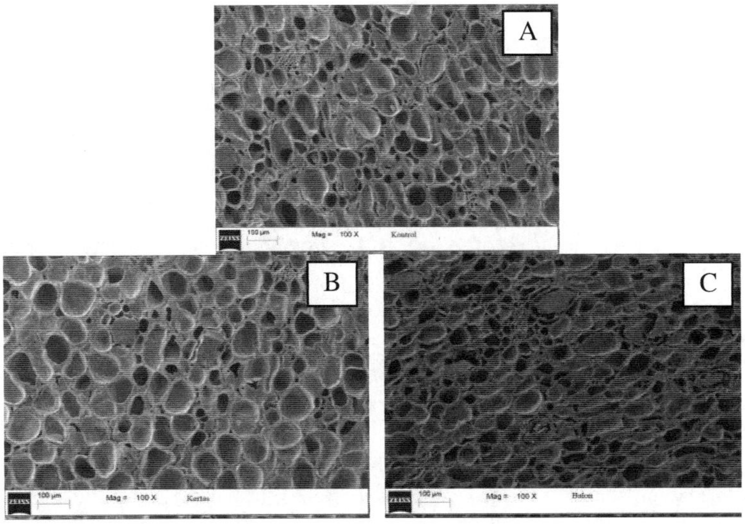


**Figure 2.** A transverse observation of mangosteen mesocarp unbagged (A), bagged with banana 'Cavendish'- paper bag (B) and bagged with baloon (C) under *scanning electron microscope* (SEM).

## 4. Conclusion

The results showed that except α-mangosteen content that was slightly decreased during the latest periode of fruit growth by bagging at preharvest, flower baggings of both bagging materials and application periods mostly did not affect mangosteen fruit qualities at harvest, but they affected fruit shelf-life and qualities during storage. Flower baggings resulted in increased fruit shelf-life, with paper bagging applied in 2 WAA was better than that applied in 4 WAA. Paper bagging in 2 WAA resulted in the mangosteen fruit shelf-life of 29 days compared to 4 WAA which resulted in 14 days shelf-life. This research proved also that the occurence of yellow latex was much more likely affected by physiological causes, not by insect attacts.

**Acknowledgments**
A special thank was directed to the General Directorate of Research Empower and Development, the Ministry of Research, Technology, and Higher Education, the Republic of Indonesia for funding this research through the National Research Grand of The Competency-based Research 2016-2018. Thanks to Great Giant Foods, Co. Ltd., Terbanggi Besar, Central Lampung through Nusantara Tropical Farm, Co. Ltd., Labuhan Ratu, East Lampung, Indonesia for providing banana 'Cavendish'-paper bags, and to Dr. Dwi Hapsoro for warm discussion during research report and manuscript preparations.

**References**
[1] Lim T K 2012 *Edible Medicinal and Non-Medicinal Plants*. Vol 2: Fruits. (Springer, New York). 1100 pp
[2] Manurakchinakorn S, Nuymark P, Phoopouk P, Poohern P and Chamnan U 2005 *Acta Hort.* **1682** 1811–18
[3] Ngarmsak M 2007 *Acta Hort.* **746** 409–15
[4] Arah I K, Amaglo H, Kumah E K and Ofori H 2015 *Int. J. Agron.* **2015** 1–6
[5] Jarimopas B, Pushpariksha P and Singh S P 2009 *Int. J. Food Prop.* **12** 414–26

[6] Cunha B L A, de França J P, Moraes A A F S, Chaves A L F, Gaiba S, Montana R, de Sacramento C K, Ferreira L M, de França L P 2014 *Acta Cirúrgica Brasileira* **29(2)** 21–28
[7] Dorly, Tjitrosemito S, Poerwanto R and Juliarni 2008 *Hayati J. Biosci.* **15** 99–104
[8] Dorly, Tjitrosemito S, da Silva J A T, Poerwanto R, Efendi D and Barasa F 2011 J. *Fruit and Ornamental Plant Res.* **19(2)** 51–65
[9] Pludbuntong W, Makhonpas C and Poovarodom S 2007 *Proc. Int. Conf. on Integ. Sci. Tech. Sustain. Dev.* 26-27 April, 2007, Bangkok, Thailand pp 30–34
[10] Poovarodom S 2010 *Proc. the 16th Asian Agric. Symp. and 1st Int. Symp. Agric. Tech.* 15–17 August 2010, Bangkok, Thailand pp 58–62
[11] Rai I N, Semarajaya C G A and Wiraatmaja I W 2011 *The Excellence Research Universitas Udayana* pp 173–78
[12] Jaritngam R, Limsakul C, Sdoodee S, Jaritngam S and Mani M 2001 *J. Agric. Sci.* **32** 1–4
[13] Jaritngam R, Limsakul C and Wongkittiserksa B 2012 *Innovative Systems Design and Engineering* **3(5)** 33–40
[14] Yantarasri T, Sivasomboon C, Uthaibutra J and Sornsrivichai J 1996 *Acta Hort.* **464** 97–101
[15] Pludbuntong W and Poovarodom S 2013 *Act Hort.* **984** 415–20
[16] Candra D, Sutikno A and Salbiah D 2013. *Pest Tropical J.* **1(2)** 1–11 (Indonesian with English Abstract)
[17] Haldankar P M, Parulekar Y R, Kireeti A, Kad M S, Shinde S M and Lawande K E 2015 *Plant Stud. J.* **4(2)** 12–20
[18] Kireeti A, Haldankar P M and Parulekar Y R 2016 *The Asian J. Hort.* **11(1)** 58–62
[19] Kurniawati A, Kasutjianingati and Bahrir M 2011 *National Seminar of PERHORTI, Lembang 23-24 November 2011* pp 1020–28 (Indonesian with English Abstract)
[20] Kurniawati A, Poerwanto R, Sobir, Efendi D and Cahyana H 2011 *J. Agron. Indonesia* **39(3)** 188-192 (Indonesian with English Abstract)
[21] Muchtaridi, Suryani D, Qosim W A and Saptarini N M 2016 *Int. J. Pharmacy and Pharmaceutical Sci.* **8(8)** 232–36.
[22] Borror D J, Triplehorn C A and Johnson N F 1996 *Pengenalan Pelajaran Serangga*. Edition 6. Partosoedjono S (translator). Brotowidjoyo M D (editor). (Yogyakarta: UGM Press) 1083 pp
[23] Palapol Y, Ketsa S, Stevenson D, Cooney J M, Allan A C, Ferguson I B 2009 *Postharvest Biol. Technol.* **51(3)** 349–53
[24] Tran D H, Yen C R and Chen Y K H 2015 *Biol. Agric. Hort. J.* **31(3)** 158–166
[25] Rai I N, Wiraatmaja I W, Semarajaya C G A, Arsana D I G K and Astiari A N K 2014 *Jurnal Hortikultura.* **24(4)** 3017-315 (Indonesian with English Abstract)
[26] Ratnamarno S, Uthaibutra J and Saengnil K 2015 *Songklanakarin J. Sci. Technol.* **27(4)** 711–717

# Analysis of Lombok 2018 earthquake effects on surface deformation process based on time lapse microgravity data (case study: east Lombok and central Lombok regencies)

**S Minardi[1,*], T Ardianto[1], A T Alaydrus[1], and M Sarkowi[2]**

[1]Department of Physics, University of Mataram, Jl. Majapahit No. 62 Mataram 83125, Indonesia
[2]Department Geophysics Engineering, Faculty of Engineering, Universitas Lampung, Jl. Prof. Soemantri Brojonegoro No. 1 Bandar Lampung 35143, Indonesia

*Email: suhayat.minardi@unram.ac.id

**Abstract:** Fault activities was detected during periods of 2016 to 2018 which presented by land subsidence in East Lombok and Central Lombok regencies during this period. In order to find out the impact of the earthquake which occurred from July to August 2018, time lapse microgravity anomaly before and after earthquake was compared, i.e. time lapse microgravity anomaly 2016 to 2018 period and 2018 to 2019 period. Measurements were carried out at about 80 points which distributed in both of regencies. Time lapse microgravity anomalies periods of 2016 and 2018 show positive values at all measurement points and deformation analysis revealed that subsidence occurred at all study areas. The results of the time lapse microgravity anomaly data processing between the periods of 2018 to 2019 show the opposite one. The results of data analysis and processing of time lapse microgravity anomaly during period 2016 to 2019 shows the deformation process on the surface before the earthquake Lombok and it was stopped due to the events of the Lombok Earthquake in 2018. The earthquake that generated by the thrust fault mechanism causing elevation change of the research area in Central Lombok Regency and East Lombok, an event of uplifting occurred.

**Keywords**: Lombok earthquake, subsidence, thrust fault mechanism, uplifting

## 1. Introduction

Research on fault activities on the island of Lombok has been carried out since 2016. During the period 2016 to 2018 using the microgravity and GPS method, fault activity has been detected [1]. The fault activity is detected signed with the changes in value of observed gravity (Gobs) during the time interval, here after referred as time lapse microgravity anomaly, and the vector shift of GPS measuring points.

Phenomena that occur due to the fault activity of one of them is the case of land subsidence in the area of research [2]. Time lapse microgravity anomaly is superposition of anomaly due to shallow and deep sources as shown by equation (1) [3]

$$\left(g_{obs(1)} - g_{obs(2)}\right) = G\int_{0}^{\infty}\int_{-\infty}^{\infty}\int_{-\infty}^{\infty} \frac{\Delta\rho(\alpha,\beta,\gamma,\Delta t)}{\left[(x-\alpha)^2+(y-\beta)^2+(z-\gamma)^2\right]^{3/2}} d\alpha d\beta d\gamma + (c_1 - c_2\rho)\Delta h \qquad (1)$$

Where, $g_{obs(1)}$ and $g_{obs(2)}$ are the gravity observation values of the first and second measurement, G is universal gravity constant, Δρ is density contrast value , (α, β, γ) and (x, y, z) are coordinate of anomaly body and measurement point respectively, Δt is time interval between first and second measurement, $c_1$ is free air anomaly constant (=0,30876 $s^{-2}$), $c_2$ is bouguer correction constant (=0,04193 $cm^3.gr.s^{-2}$), and Δh is the elevation change of the first and second measurements. Anomaly due to deep source is shown by the first term of right segment of equation (1) is caused by subsurface dynamics and anomaly due to shallow source is shown by second term is caused by vertical deformation on the surface called a land subsidence.

Land subsidence could effort by many factors i.e. groundwater extraction, induced by the load of constructions, natural consolidation of alluvial soil, and tectonic activities [4]. In Lombok Island natural consolidation and tectonics activities are significant factors affected the land subsidence. To get an information about land subsidence, value of time lapse anomaly due to shallow source has to separate from total time lapse anomaly.

Method of anomaly separation that used in this research is striping filter, which need an input thickness, depth, and contras density of shallow and deep layer [5, 6]. To extract shallow anomaly spectrum, filter stripping for a can be written as:

$$S(u,v) = \left(1 + \alpha\beta(u,v)e^{-k\xi}\right)^{-1} \qquad (2)$$

Where, $\alpha = \frac{\Delta\rho_d}{\Delta\rho_s}$, $\beta = \frac{1-e^{-kt_d}}{1-e^{-kt_s}}$, and $\xi = h_{t_d} - h_{t_s}$, which α is density comparison, β is thickness comparison (t) and ξ is the difference of depth between the shallow and deep layers.

In the period of July to August 2018 Lombok is shocked by at least 6 earthquakes with a magnitude of more than 5 Richter scale (RS) as shown in **table 1**.

**Table 1.** Parameters of Lombok earthquakes 2018 which has magnitude more than 5.0 RS [7]

| Date of event EQ | Magnitude (RS) | Position of hypocentre | Depth of EQ source (km) |
|---|---|---|---|
| 29 July 2018 | 6.4 | 116.508E, 8.324S | 14.0 |
| 05 August 2018 | 6.9 | 116.436E, 8.260S | 31.0 |
| 09 August 2018 | 5.9 | 116.208E, 8.394S | 10.0 |
| 19 August 2018 | 6.3 | 116.576E, 8.325S | 7.9 |
| 19 August 2018 | 6.9 | 116.626E, 8.324S | 25.0 |
| 25 August 2018 | 5.5 | 116.425E, 8.425S | 9.2 |

Based on the location and depth of the hypocentre, this earthquake is a type of shallow earthquake due to the Fault Activity of the Flores Back Arc Thrust. The results of the analysis of the source mechanism show that this earthquake is generated by rock deformation with a thrust fault mechanism [8]. These earthquakes will affect the deformation process which has been observed before.

## 2. Methods

Research was done at area about (50 x 30) km2, with a measurement station of 80 units spread in central Lombok and East Lombok Regencies as shown in figure 1. Measurements were made 3 times, in August 2016, April 2018, and June 2019 using gravity meter Scintrex Autograv CG-5 as shown in figure 1 and figure 2.

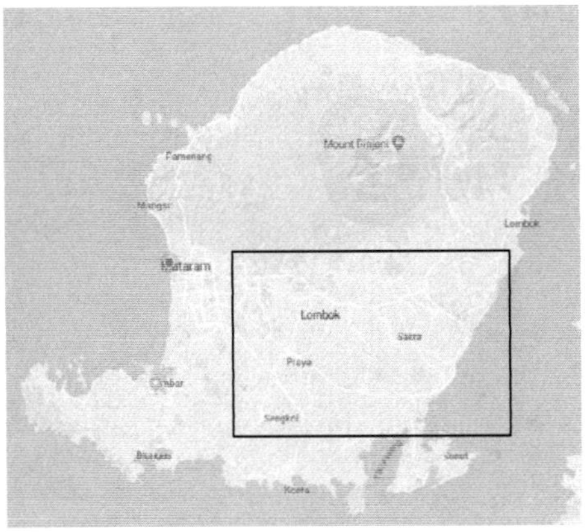

**Figure 1.** Data acquisition location (unscaled) [9]

**Figure 2.** Scintrex Autograv CG-5 gravimeter [10]

Surface deformation analysis using striping filter with input in the form of subsurface structure analysis based on gravity data in 2016. Analysis was carried out on time lapse microgravity anomalies data periods 2016 - 2018, 2018 - 2019, and 2016 - 2019. The stage of processing data was shown in **figure 3.**

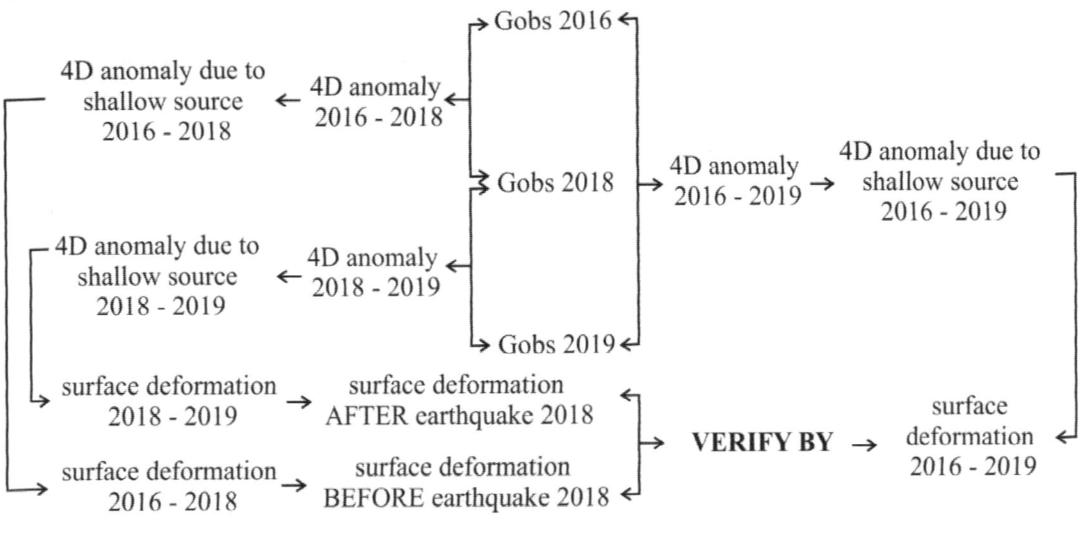


**Figure 3.** Stages of processing data to analyse an effect Lombok Earthquake 2018 on surface deformation

## 3. Results and Discussions

Time lapse microgravity anomalies periods of 2016 and 2018 show positive values at all measurement points. The results of time lapse anomaly separation show that time lapse anomaly due to shallow source is about (6 – 30) □Gal. Conversion the anomaly resulting surface deformation as land subsidence of 2 to 10 centimetres occurred at all study areas as shown in figure 4. Land subsidence that occurred due to natural consolidation, geologically research area composed of young rocks (Quaternary aged), and triggered by tectonics activity of Lombok Island.

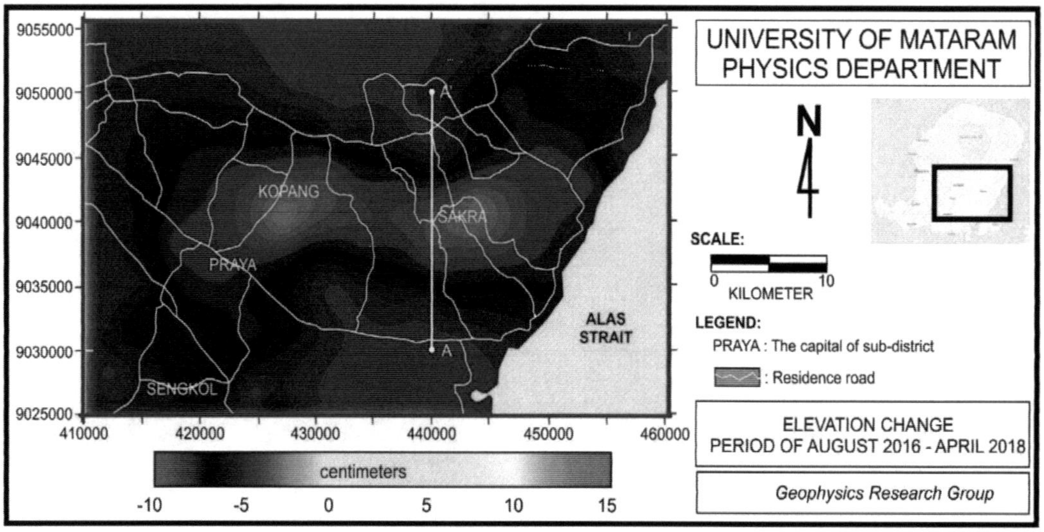


**Figure 4.** Elevation change period of August 2016 to April 2018

The time lapse microgravity anomaly data processing between the periods of 2018 to 2019 show the opposite one, most of the study areas have negative anomalous due to the shallow source. Negative change values, which mean Gobs which measured in 2019 are less than Gobs 2018. The results of surface deformation analysis show that uplifts happened at all study areas ranging from 0.5 to 13 cm as shown in figure 5.

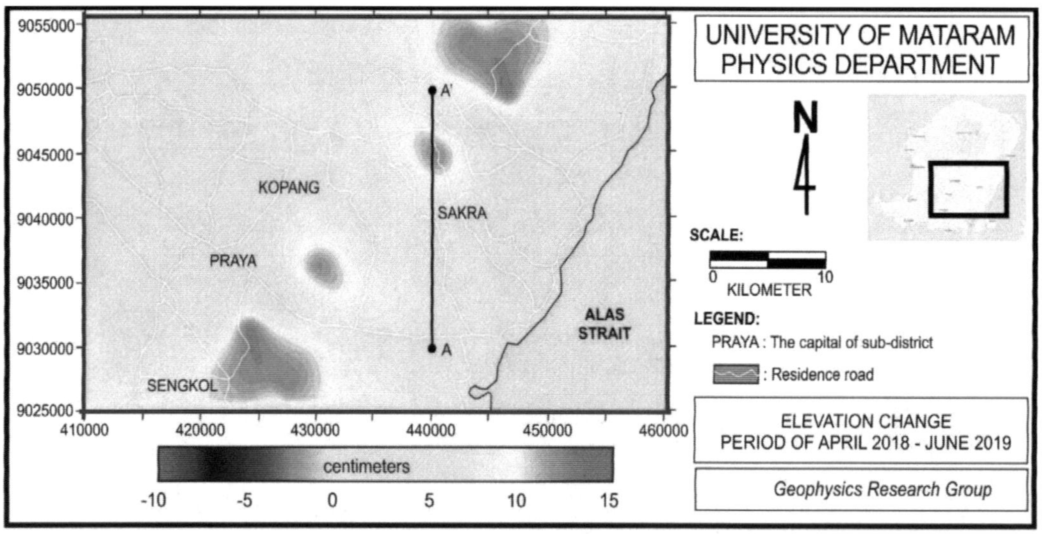


**Figure 5.** Elevation change period of April 2018 to June 2019

Change of Gobs value along period of 2016 to 2019 is superposition of change values along period of 2016 to 2018 and 2018 to 2019. Time lapse anomaly was founded due to shallow source in this period was -26 to 14 μGal and this value is an anomaly due to the elevation change as much ass -5 to 9cm, refer to fig. 6.

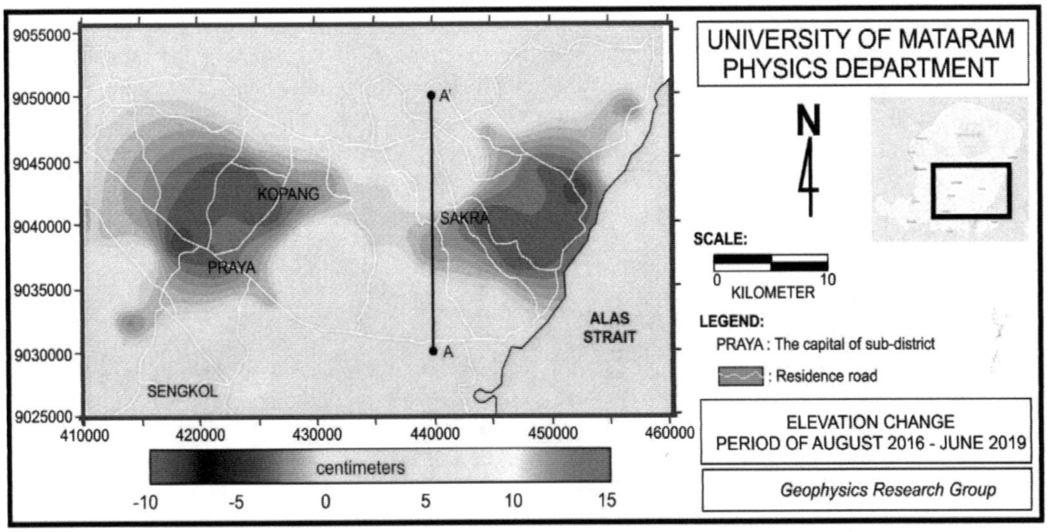


**Figure 6.** Elevation change period of August 2016 to June 2019

To see the impact of the 2018 earthquake also made a cross section AA' as shown in figures 4, 5, 6. Time lapse anomalies along period of 2016-2018 in AA' cross section are positive, shown in figure 7, as well as the results of filtering produce anomaly due to shallow sources is also positive. This is because land subsidence (change in negative elevation), shown in figure 8, results in Gobs 2018 value greater than 2016.

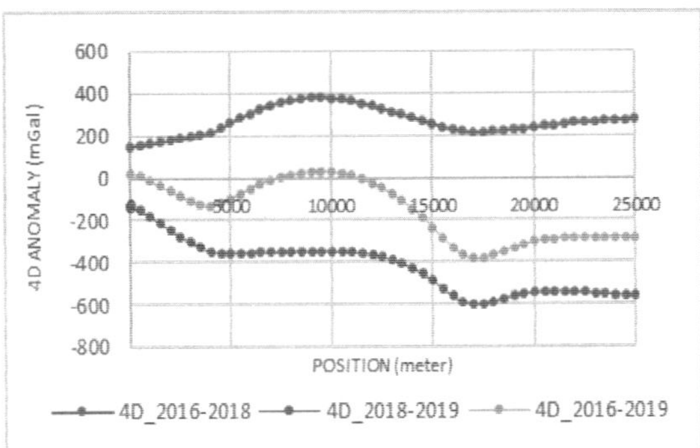


**Figure 7.** Time lapse (4D) microgravity anomaly periods of 2016 to 2018, 2018 to 2019, and 2016 to 2019 along cross section AA'

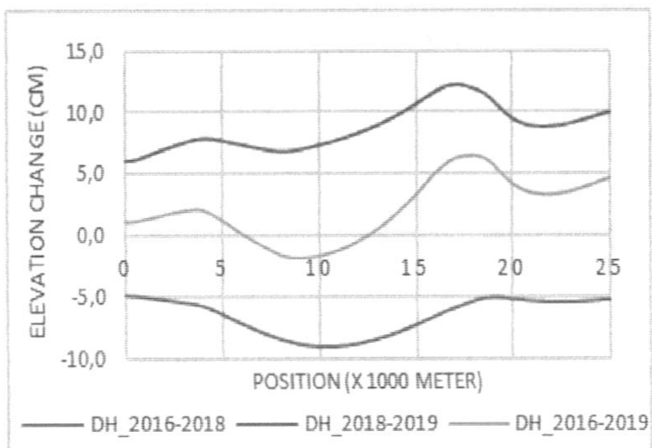


**Figure 8.** Elevation change periods of 2016 to 2018, 2018 to 2019, and 2016 to 2019 along cross section AA'

Based on figure 7, 8 and also figure 5, in the period 2018 - 2019 the opposite situation occurs. The time lapse anomaly value in this period is negative. The results of filtering with striping filters produce negative values as well, indicating that on the cross-section AA', and most areas of the study, an uplifting process occurs.

Change of elevation along periods 2018 to 2019, which dominated by uplifting process as impact of earthquake not able to restore the state to the condition before the earthquake. This can be seen in Figure 8 that the change in elevation along period of 2016 to 2019, in cross section AA still leaves a negative elevation changes (subsidence) in some places.

The earthquakes as stated in table 1 are all centred north of Mount Rinjani, and with regard to their location and hypocentre depth, this earthquake is a type of shallow earthquake due to the Fault of Back Arc Thrust activity. Results of the analysis show the mechanism of the source of these earthquakes, earthquakes generated by the deformation of the rock with the upward movement mechanism (thrust faults). The mechanism of the upward movement in the northern part of Lombok Island will affect the southern part which is also somewhat pushed up by the Indo Australia plate. The impact pressure from the north and south of the island of Lombok looked forming anticline structure, with a peak around Praya, Kopang, and Sakra, as shown in Figure 6. The uplifting process due to tectonic activity (earthquake) was impact natural deformation on the surface, which is land subsidence.

## 4. Conclusion

Lombok earthquakes on July to August 2018 significantly affected to deformation on the surface. Surface deformation as land subsidence which happen before earthquake occurred, was stopped due to earthquake and become an uplifting.

**Acknowledgments**

This research is funded by Directorate of Research and Community Service, Directorate General of Strengthening Research and Development at the Ministry of Research, Technology and Higher Education with contract number: 068/SP2H/LT/DPRM/2019. We also thank to the chairman and all staff research institutions of University of Mataram, Meteorologically, Climatologically, and Geophysical Agency (BMKG) and operator who provide to the gravimeter during the research.

## References

[1] Minardi S, Ardianto T and Alaydrus A T 2019 Deteksi Aktivitas Sesar Lombok Berdasarkan Data Anomali Gayaberat Antar Waktu Dan Geodetik, *Proc. Seminar Kontribusi Fisika 2018* (Bandung: Depertemen Fisika ITB)

[2] Minardi S, Ardianto T, Alaydrus A T 2019 Analyzes Of Subsidence in Central and East Lombok Districts Using Time Lapse Microgravity Data Periods of August 2016 to April 2018 *AIP Conference Series (New York: AIP)*

[3] Darharta D, Sarkowi M, Kadir W G A and Minardi S 2007 Penurunan Volume Airtanah Daerah Semarang berdasarkan Pemodelan 3D Gayaberat Antar Waktu *J. Geoaplika* (Bandung: ITB), **2**, No. 1, pp. 11 – 17

[4] Abidin H Z, Andreas H, Gumilar I, Fukuda Y, Pohan Y E and Deguchi T 2011 Land subsidence of Jakarta (Indonesia) and its relation with urban development *Natural Hazards* **59 3**

[5] Cordell L 1985 A stripping filter for potential-field data *55th Annual International Meeting, SEG, Expanded Abstracts*, 217-218

[6] Aina A 1994 A Simple Of Cordell's Stripping Filter *Geophysics* **59** 488-490

[7] https://regional.kompas.com/read/2018/09/23/11321551/melihat-kembali-gempa-lombok-2018-dan-sejarah-kegempaannya?page=all. downloaded in June 16th, 2019 at 5:12 PM

[8] https://id.wikipedia.org/wiki/Gempa_bumi_Lombok_5_Agustus_2018, downloaded in July 6th, 2019 at 7:12 AM

[9] https://www.google.co.id/maps/@-8.6071255,116.071663,10z?hl=en&authuser=0, downloaded in December 16th, 2018 at 5:12 PM

[10] https://www.geomatrix.co.uk/land-products/gravity/scintrex-cg5, downloaded in December 16th, 2018 at 5:20 PM.

# The potential of Lampung province as the area for producing mineral fertilizer

**K Isnugroho[1], Y Hendronursito[1], D C Birawidha[1], M Amin[1], M A Muttaqi[1],**

[1] Research Unit for Mineral Technology, Indonesian Institute of Sciences, Tanjung Bintang, Lampung, Indonesia

*Email: kusn005@lipi.go.id

**Abstract**. Lampung Province has the potential as a producer of mineral fertilizers. The availability of mineral and rock resources in the area has the potential to be processed into mineral fertilizers. Rocks and minerals have been shown to increase productivity and maintain fertility of agricultural land. Rocks and minerals such as zeolite, basalt, pumice, limestone, phosphate and pearlite, its availability in Lampung Province makes it possible to become an agro-mineral industry raw material. For this reason, there is a need for good cooperation between related agencies to create processing of rocks and minerals into effective, efficient mineral as fertilizers.

**Keywords**: Rocks, minerals, fertilizers, agro-mineral, agricultural

## 1. Introduction

The availability of food is necessary to compensate for the increase of the population. The impact of increasing population land use has shifted. The total of agricultural land continues to decrease because the function of land is transformed into residential land. The decreasing of soil quality in agricultural land causes the decreasing in productivity. To overcome this problem, there are two things. First, control the conversion of agricultural land. Second, improve the quality of critical land to function as agricultural land [1]. To maintain and increase the productivity of the soil, it is necessary to add nutrients by fertilizing. Unfortunately, the chemical fertilization process will damage the soil structure itself. Adding minerals into the soil that a good technique to restore the soil fertility [2]. Fertilizers arematerials that contain one or more nutrients needed by plants including C, H, O, N, P, K, Ca, Mg, S, Fe, Mn, U, Zn, L, Mo, and B.

The important production factor for the agricultural sector is fertilizer. Fertilizers contribute to the success of increasing agricultural production by 20 percent [3]. The most essential plant nutrients are nitrogen (N), phosphorus (P) and potassium (K). Except for some special nitrogen fertilizers, almost all artificial fertilizers from rocks which chemically processed, i.e. rocks that have been chemically modified, plus some micronutrients that are often needed by plants, such as calcium (Ca), magnesium (Mg), sulfur (S), copper (Cu), cobalt (Co), iron (Fe), and etc. Generally, only modified physically, for example by crushing. However, modifications is needed for rocks and minerals (derivative), which has a number of specific chemical substances combined with an aggregator, but only simple techniques consumption that can be optimized.Therefore, agro-mineral is expected to be an alternative substitute for fertilizer that is cheaper and easier to obtain to increase plant nutrition and improve soil structure, by utilizing the geological resources found around the agricultural land [4].

Natural minerals contain many macro and micro elements which can be used as nutrients for plants. The macro and micro elements in natural minerals depend on the building blocks of rocks. The application of natural minerals as a source of nutrients is a natural process that utilization for fertilizer will be safe for the environment [3].

Lampung Province has the potential of natural resources in the form of metallic, industrial, energyand construction minerals. Nevertheless, the findings related to mineral deposits in the area of Lampung has not been found, both in quality and quantity. In this paper, the authors try to explain the potential of mineral deposits in Lampung province whichcan be used to build the mineral fertilizer industry.

## 2. Description of Lampung Province

Lampung province is an area with average high around 300-500 m of the sea. It is located between $103^0$ 40'and $105^0$50' east longitude, $6^0$45', and $3^0$45' South latitude. Lampung province area is shaped by 34.623.80 $km^2$. In terms of geographic position has Lampung province boundaries as follows: North-South Sumatra province and Bengkulu province; South- Sunda Strait; West Indonesia Ocean; East: Java Ocean. Lampung province has 13 regency and 2 municipality, include;West Lampung, Tanggamus, East Lampung, South Lampung, North Lampung, Way Kanan, Tulang Bawang, Pesawaran, Mesuji, Pringsewu, East Tulang Bawang, The West Coast regency, Bandar Lampung, and Metro municipality[33].

### *2.1 Geology Lampung Province*

Lampung province has a total area of 35288.35 $Km^2$ including the islands located at the most south-eastern part of SumateraIsland. Lampung province is bordered by the province of South Sumatra and Bengkulu to the North, The Sunda Strait to the South, the Java Sea to the East, the Indonesian Ocean to the West. Western ridge of Lampung is part of the Bukit Barisan mountain range of the which is Geanticlinal and Synclinal to the east. This part of the mountains from the Cretaceous age, underwent deformation process during the tertiary age to form a fault. This process generates a geological phenomenon like long-cal Watermelon fault along the river of Way Watermelon and gulf of Watermelon and oval-shaped volcanoes (Tanggamus, chills, Rebang surround it). Tectonic depressions such as the valleys of Suoh, Gedong Surinam, and Way Lima roommates were covered by Volcanic sedimentation from the fisuves eruption. Sukadana Basalt, a plateau, was the result of an eruption that is happened in the Holazin age. This plateau was not sediment like volcanic sand resulted from volcanic dust. Data on the mineral deposits in Lampung are not available yet so that the potential of mineral deposits is not much known [33].

### *2.2 Agricultural conditions*

Wetland paddy production in Lampung province Reached 4.09 million tons during 2017 Increased by 6.75 percent, the highest production generated by the Central Lampung Regency, roommates Reached 733.03 thousand tons. The highest productivity of rice crops in Bandar Lampung City is 59.86 quintal/hectare. Horticultural type of plant for vegetable crops, the Reviews largest production produced by the plant petsai roommates Reached 10 684 tonnes, where 53.04 percent is produced from West Lampung Regency. While for the type of fruit plants produced the Reviews largest production of banana fruit roommates Reached 1462.42 thousand tonnes with 32.98 percent produced from Central Lampung Regency. Lampung province is famous for the production of coconut, rubber, and coffee, it is supported by both kinds of production of Reviews These crops. In 2017, Lampung is Able to produce 170. 475 tonnes of oil palm and 114 214 tons of coffee. The production of palm oil produced by Central Lampung Regency roommates Reached 25.38 percent of the total production, while coffee production ranks third after rubber production. The largest coffee produced from West Lampung area Reaches production of 50.35 percent [33].

## 3. Discussion

In Lampung province, various mineral resources are founded. There are several minerals and rocks that have the potential to be used as raw materials for the mineral fertilizer industry including:

### *3.1. Zeolite*

Lampung has abundant natural zeolite resources, spread in several districts in Lampung province. Data from the Directorate of Regional Potential Development, in 2012, Lampung had a source of natural zeolite of 31.173.505 tons [5]. Meanwhile, according to Kusdarto, the availability of zeolite in Lampungas much as 200.000 tons in Kalianda District, and 2.000.000 tons in Katibung District both in South Lampung Regency. While in Tanggamus District, Cukuh Balah Sub-district there are 41.600.000 tons of natural zeolite. Zeolite rocks in the Southern District are white tufa, fine-grained to coarse. Mineral consists of clinoptilolite and mordenitewith a mineral association of plagioclase, montmorillonite, cristobalite, and quartz. Even though zeolite from Tanggamus is greenish white, hard, and massive [6]. Research and application of zeolite in agriculture has developed. Zeolite has been proven to help the process of iincreasing production in agriculture, improving the nature of the soil which is physically and chemically damaged.

The function of zeolite in this case as a soil conditioner, carrier of fertilizer, controlling the release of ions $NH4^+$ and $K^+$ (as a slow release fertilizer) and maintaining moisture of soil. Natural zeolite has a good ability to absorb and exchange cations [6-7].Zeolite is not acidic, so it can support the acidity of the soil and reduce the dose of lime [8]. Five types of zeolite which have been beneficial for agriculture such as clinoptilolite, mordenite, erionite, kabasit, and philipsit. The result analysis of soil showed that zeolite was increased the content of K, Na, Ca, and Mg soil. It was also increase the Cation Exchange Capacity (CEC) and soil pHbecause the cations in zeolite were pushed out by $H^+$ and the cation was released into the soil solution which may cause the nutrient supply [7]. The application of zeolite has been carried out in various types of soil with various types of plantsand showed a positive result to increased productivity and soil structure [9-13].

### *3.2. Pumice*

In Lampung province, data related to pumice resources has not been widely published. However, based on a map of the potential of non-metallic mineral resources in Lampung province, issued by the Directorate Inventory of Mineral Resources, there is potential resources of pumice in South Lampung, Kalianda district. Pumice is a volcanic rock with light colored, chemically and physically unobtrusive, like perlite.

These rocks are formed as the result of a great expansion of the gases that are dissolved in a silica-rich lavas such as rhyolite or riodasit thick. Like perlite, pumice has pores, so it can float on water. For horticulture, pumice was took from loose sediment with splitting and sorting. For the environment, the production of pumice more friendly than perlite or vermiculite because it does not require high energy for thermal expansion. The difference with perlite mainly related to pore size, particle shape, and size [4]. Pumice has the physicochemical properties that are similar to perlite so it can be used to replace perlite which is not expensive in greenhouse plant maintenance [4].

### *3.3. Limestone*

Everywhere the soil acidity is a common problem due to the decreasing in agricultural products, including developed countries. High solubility rates, unsuitable rock, and some things are rare such as continuous chemical acidification and the addition of large amounts of ammonium sulfate. The main problem in the highly acidic farming area is insufficient high $H^+$ ions,the addition of highly toxic $Al^{3+}$ concentrations at low pH, and low levels of $Ca^{2+}$ ion exchange. These issues are generallyhandled by liming. All materials containing Ca and Mg compounds can be used as liming materials to neutralize the acidity of the soil, increasing the pH of soil with adding of Ca and decrease of Al.

Limestone and dolomite are materials that have been used for liming for centuries, and it still used in some countries [4]. The acidity of soil have physical and chemical constraints that inhibit the growth of plants. Fertilizing and liming of acid soils can make the land productive. Lime is a group carbonates such as calcite ($CaCO_3$) and dolomite ($CaMg(CO_3)_2$), commonly used to increase the pH of soil because it will dissociate into $Ca^{2+}$ and $Mg^{2+}$ ions in the soil [32].

In Lampung province, there are 1.900.000 tons of limestone, spread in Pringsewu, Pesawaran, and North Lampung. Limestone is a type of carbonate rock that occurs in nature, also called limestone. The main mineral of limestone is calcite ($CaCO_3$), other minerals are impurity minerals, usually consisting of quartz

($SiO_2$), carbonate associated with iron minerals and clay minerals, as well as the remaining organic matter of plants [15].

### *3.4. Perlite*

Perlite is a term for volcanic glass material which is not processed and processed.Raw perlites which are not processed are metastable, amorphous, silica-rich volcanic rocks which have a riodytic and riolytic composition. It colored light gray, black glass, and has several concentric fractures resembling the skin of an onion. When heated at a temperature of 1000 °C,the rock will expand to 20 times the volume initially by evaporation of water trapped, white, glass foam forming commercial perlite, porous, lightweight, sterile, physically silicate stable with a good properties heat-retaining (good thermal insulation), and neutral pH.

The expanding pearlite has a low bulk density but practically not having cation exchange properties [4]. Perlite may not be directly related to the soil or plants, but widely used as a aggregate of horticulture. The properties of perlite, pH between of 6.5-8.0, and the cell structure causes perlite to absorb the solution many times (can carry herbicides, insecticides, fertilizers, soil regulators and plant roots) [2]. The potential of Perlite in Lampung province was found in West Lampung and North Lampung, with the total mortgage amount of 79 million tons.

### *3.5. Basalt*

The potential of basalt rocks in Lampung province, spread around the mountain from the west side Mt. Tanggamus, through around the semangko bay and continues to the east of Mt. Rajabasa and basalt plains in the Sukadana area. The distribution of basalt divided into two, western volcanic rocks (Oligo to Hulusimpang Miocene formation) and eastern volcanic rocks (from Sukadana to Tamiyang). The difference between eastern and western of basalt in the content of MgO and CaO. Eastern basalt high MgO (7wt%) and low CaO (8wt%), while western basalt low MgO (4wt%) and high CaO (9wt%) [16].

Basalt potential reached 419.071.833 $m^3$ spread over various districts in Lampung province.Some research and application of basalt rocks as mineral fertilizer materials have been carried out [17-22]. The technique of using minerals as soil fertilizer canprovide the supply of macronutrients such as nitrogen, phosphorus, potassium, calcium, magnesium, and sulfur,as well as micronutrients such as Fe, Mn, U, Zn and Na on the surface of the ground,this technique also called stone meal fertilizer.Basalt rock can be used as a fertilizer with combined or mixture with other ingredients [21]. In addition, the application of basalt rocks shown an increasing of productivity and growth of plants [23]. Several studies show the benefits of the application of basalt to improve pH, as an alternative to reduce costs and imported raw materials while avoiding negative environmental impacts, improving the Cation Exchange (CEC) of land, and adding K elements to the soil [19] [20] [24-25].

### *3.6. Phosphate*

Natural phosphate (rock phosphate) is the common name used for some types of rocks that contain amounts of phosphate mineralsor contains phosphate ions in their chemical structure.Many types of rocks have phosphate-containing components, but rocks containing amount of phosphate that has economic value as mineral or ore is not prevalent [26]. The definition of natural phosphate according to the American Geological Institute is sedimentary rock composed mainly of phosphate minerals [27]. Based on the mineral composition of phosphate sedimentary rock can be distinguishedinto phosphate-Ca, phosphate Ca-Al-Fe and phosphateFe-Al [28].

Natural phosphate has high solubility in acidic conditions, therefore it is very suitable when used as a source of P fertilizer on dry land such as Ultisol, Oxisol, and some Inceptisol, andless suitable for use in neutral and alkaline soil reaction [26]. Natural phosphate can be used directly as fertilizer and it known as effective as liquid P fertilizer in several countries [29]. The application of natural phosphate from North Carolina and Tunisia in Red Yellow Podzolic soil with P deficiency, pH of 4.5, and saturation aluminum of 78% for fourconsecutive planting season for upland rice and corn crops showed thatthe natural phosphate was more effective than TSP [30].

Other studies on podzolic soil from Lampung and Jambi showed that natural phosphate has the same effectiveness and even better than TSP [31]. In Lampung province, phosphate reserves has not been recorded. However, based on the map of Lampung province, the potential of non-metal mineral resources from the Directorate Inventory of Mineral Resources, there is potential for phosphate in North Lampung.

## 4. Conclusion

The potential of mineral resources in Lampung province needs to be further investigated in terms of quality and quantity. Zeolite, pumice, limestone, pearlite, phosphate, and basalt can be an alternative sources to supply of mineral fertilizers. The good cooperation between the local government and research institutions (Indonesian Institute of Agricultural Technology, Indonesian Institute of Sciences, and University) in Lampung province needs to be developed for agro-mineral industry. The above description about Lampung province provide advantage for developing of mineral fertilizer industry. As the gateway island of Sumatera, and supported by the International class port, Port of Panjang, the mobility of the product will be easy to distribute to the various country in Indonesia. It can be seen the existing of agricultural conditions in Lampung province, it shown the potential market that can be absorbed by producers of mineral fertilizer.

## Acknowledgments

The authors would like gratefully acknowledge to the research program of Research Unit for Mineral Technology-Indonesian Institute of Sciences (BPTM-LIPI), independent research, and Insinas 2019 for the facilities and research grant supporting in this research.

## References

[1] Hermawan B, 2011 *Proc. Seminar Nasional BudidayaPertanian*: 60-70.
[2] Mawarda I W, Titisari A D 2004 Agromineral (Mineralogi Untuk Ilmu Pertanian) Yogyakarta: UGM.
[3] Sembodo S, Noviardi R, Suyadi D, Hidayat E 2013 *Proc. Pemaparan Hasil Penelitian Puslit Geologi – LIPI.*
[4] Kusdarto 2005 Potensi Agromineral di Indonesia Salah Satu Alternatif Pengganti Pupuk Buatan, Sub-Dit Mineral Non Logam Kementerian Energi Sumber Daya Mineral.
[5] Darmansyah, Ginting S, Widiana D, Saputra H 2013 *Proc. Pemaparan Hasil Penelitian Puslit Geologi – LIPI.*
[6] Kusdarto 2008 Potensi Zeolite di Indonesia, *J. Zeolit Indonesia* **7** (2): 78-68.
[7] Sabilu Y 2106 *Biowallacea* **3** (2): 396-407.
[8] Fuji S 1974 Heavy Metal Adsorption by Pulverized Zeolites, Japan, Kokai 74:849.
[9] Abadzic S D, Ryan J N 2001 *J. Environmetal Science and Technology*, **35**: 4501-4508.
[10] Kavoosi M 2007 Communication in Soil Science and Plant.Analysis 38: 69-76.
[11] Abdi G H, Khui M K, Eshghi S 2006 *Int. J. of Agricultural Research* **1**: 384-389.
[12] Ando H, Mihara C, Kakuda K, Wada G 1996 Soil Science and Plant Nutrition 42: 531-538.
[13] Wiedenfeld B 2003 Subtropical Plant Science 55: 7-10.
[14] Noland A D,SpomerA L, Williams J D, 1992 Communication in Soil Science and Plant Analysis 23(13&14): 1533-1547.
[15] Aziz M 2010 *J Teknologi Mineral Batubara* **6**(3): 116-131.
[16] Zulkarnaen I 2011 Geochemical Evidence of Island-Arc Origin for Sumatra Island ; A New
[17] Perspective based on Volcanic Rocks in Lampung Province , Indonesia. *J. Geol. Indones.* **6**, 213–225.
[18] Bockman O C 2016 Fertilizers and biological nitrogen fixation as sources of plant nutrients: Perspectives for future agriculture Author ( s ): O.C. Bøckman Source : Plant and Soil , Vol. 194(1/2): 11–14.
[19] HarleyAD, Gilkes, RJ 2000Factors influencing the release of plant nutrient elements from silicate rock powders: a geochemical overview. NutirentCycl. Agroecosystems 56: 11–36.

[20] Nunes JMG, Kautzmann RM, Oliveira C 2014 Evaluation of the natural fertilizing potential of basalt dust wastes from the mining district of Nova Prata ( Brazil ). J. Clean. Prod: 4–
[21] Theodoro SH, Leonardos, OH 2006 The use of rocks to improve family agriculture in Brazil 78: 721–730.
[22] Shamsudin J 2016 Improving the Productivity of Acid Sulfate Soils for Rice Cultivation using Limestone, Basalt,Organic Fertilizer and/or their Combinations 45:383–392.
[23] Keller W D, 2019, Native Rocks and Minerals as Fertilizers. Sci. Mon. 66:122–130.
[24] Goreau T J, Stack E, Senechal, E Tang, JRyals, RVanacoreT, Campe J2013 Basalt Dust and Biochar Interactions at New Harmony Farm, Massachusetts 1634.
[25] Fyfe WS, Leonardos, OH, Theodoro SH 2006 Sustainable farming with native rocks : the transition without revolution 78:715–720.
[26] Wang Y, QI, X.-N 2007Evaluation on black soil nutrients and fertilization recommendation in Dehui City. Chinese J. Eco-Agriculture 15:26–28.
[27] Kasno A, Rochayati S, Prasetyo H B 2009 Fosfat Alam: Pemanfaatan Fosfat Alam yang Digunakan Langsung sebagai Pupuk Sumber P : 1-23.
[28] Gary M, R McAfee Jr, C L Walf (eds) 1974 Glossary of Geology.Amer.Geolog.Ins. Washington D.C.
[29] McClellan G H, L R. Gremillon 1980 Evaluation of phosphatic row materials. In F.E. Khasawneh, E. C. Sample and E. J. Kamprath (Eds). The role of phosphorus in Agriculture. Soc.Of Amer. Soil Sci. Soc. Of Amer. Madison, Wisconsin, USA.
[30] Engelstad O P, A Jugsujinda, S. K. De Datta 1974. Respone by flooded rice to phosphate rock varying in citrat solubility. Soil. Sci. Soc. Amer. Proc. 38: 524-529.
[31] Chien S H, L L Hammond 1978 Comparison of various laboratory methods for predicting the agronomic potential of phosphate rocks for direct application. Soil Sci. Soc. Amer.J. 42: 935-939.
[32] Sri Adiningsih, Sri Rochayati. 1990 Use of phosphate fertilizers in arable food crop productin in Indonesia. Center for Soil and Agroclimate Research, Bogor, Indonesia.
[33] Rochayati S, Adiningsih, J S, Didi Ardi S. 1986 Pengaruuh pupuk fosfat dan pengapuran terhadap hasil kedelai dan jagung pada tanah Ultisol Rangkasbitung. Pemberitaan Penelitian Tanah dan Pupuk (5) : 13-18.
[34] Lampung Province in Figure 2018 Statistics of Lampung Province

# A study on drill machining for Magnesium alloy using Taguchi method

**G A Ibrahim[1*], Y Burhanuddin[1], and D Emrijakto[1]**

[1] Mechanical Engineering Department, University of Lampung, Jl. Prof. Sumantri Brojonegoro no 1, Bandar Lampung

*Email: gusri.akhyar@eng.unila.ac.id

**Abstract.** Magnesium is one the very important material light weight, especially in the field of the vehicle component, and the field of biomedical. Application of material magnesium and its alloy in the field of biomedical based on the magnesium characteristics which look very much like the nature of biocompatibility of bones in man and melted down within the body. Application material magnesium are also many developed by means of grown (implant). Magnesium alloy material has some of the more in the chemical nature and physics, but in the machining process a magnesium alloy known as combustible metal material type, especially during the machining process at high speed. The optimizing of machining process is necessary in order to maintain for not chip burning. In this research, the point angle, lubricant and drilling parameter were analysed to know their influence on the value of surface roughness, cylindricity and perpendicularity. On this research analysis support for new carried out using a Taguchi Method L18 consisting of 3 factors with 3 levels and 1 factors with 2 levels. The research results show that surface roughness is influenced by the significant contributions from the point angle 18.9 % and lubricant 14.5 %, where the point angle is 650 with lubricant of synthetic oil. The emergence of a nose radius tendency resulting in the increasing level subtlety. Cylindricity is influenced by an point angle 450 of the drilling process however, especially if there is an interactions of the feeding of 0.2 mm/rev and synthetic lubricant. The cutting edge accuracy led to a tangential force so that had an impact on their rotation stability and the cylindricity of drilling results. The value of perpendicularity of magnesium AZ31 in the drilling process however also carried out similar procedures are very much influenced by the cutting parameter of feeding that produces the significance of p = 0,044 , while the strongest significance will be happened if there were an interaction between feeding of 0,1 mm and synthetic lubricant to p = 0.041.

**Keywords:** drilling, magnesium AZ31, surface roughness, Taguchi method, perpendicularity

## 1. Introduction

Drilling process is one of the most important cutting processes and has been widely investigated by numerous researches. In fact, drilling operation is not considered as a very crucial factor to influence the quality of machine parts with a good machinability like magnesium alloys [1]. Many researchers have investigated effects of drill bit's geometrical parameter on the quality of machine parts. Chen and Fuh [2] used a split-point drill to develop a force model incorporating the splitting parameters for the drilling process. They optimized drill-point geometry to minimize the thrust force and torque during drilling. Chen and Tsao [3] indicated that drills with different coatings performed differently on the drill life and quality of drilled hole. Haan and Batzer [4] investigated effects of different profiles on the functions of drills. They measured torque produced from varying drilling conditions: cutting speed, feed, on/off in running cutting fluid. They concluded that cutting fluid is a significant factor in producing drilling hole's profile. Besides, cutting fluid affected the amount of BUE (Built-up Edge) depending on the efficiency for cutting fluid removing heat from cutting work piece [5] used the tractable metals and developed empirical expressions for estimating torque and thrust force during drilling process. They indicated that increasing the point angle of drill reduced torque but increased the thrust force. The above information indicates that geometrical factors of drill bits and processing parameters greatly affect the performances of drilling operation.

Magnesium is one the very important material light weight, especially in the field of the vehicle component, and the field of biomedical [6]. Application of material magnesium and its alloy in the field of biomedical based on the magnesium characteristic which look very much like the nature of biocompatibility of bones in man and melted down within the body. Application material magnesium is also many developed by means of grown (implant) [7]. Magnesium alloy material has some of the more in the chemical nature and physics, but in the machining process a magnesium alloy known as combustible metal material type, especially during the machining process at high speed [8 9]. The optimizing of machining process is necessary in order to maintain for not chip burning. In this research, the point angle, lubricant and drilling parameter were analysed to know their influence on the value of surface roughness, cylindricity and perpendicularity.

However, drilling process is a necessary step for making parts. So far, most researchers are trying to prevent the ignition of magnesium scraps through modified the geometry of machining scraps and machining handbooks recommend that drill bit of 700–1180 point angle is suitable for drilling magnesium and magnesium alloys. This study, however, investigated not only the operation parameters, like cutting speed, feed rate and point angle of drill bits on the quality of drill hole, but also the dynamic effects of drilling operation on microstructure of AZ61 [1].

The experiment was carried out on the basic of Taguchi's L18 Orthogonal Array of experiments. The important input drilling parameters were chosen as spindle speed (n), feed rate (f), point angle and lubricant. In order to minimize the values of all the above-mentioned performance characteristics, an optional combination of input drilling parameter is required. Taguchi optimization technique is used to for optimization drilling parameters is required: ANOVA is used to find the highly influential drilling parameter(s) that contributes to a highly quality product.

## 2. Experimental Procedure

A vertical type of Vertical Machining Centre (VMC) was used in this study. Magnesium is metal who is lightweight (1.74g/cm3), 1.6 times lighter than al and 4.5 times lighter than steel. The toughness of broken is greater and mightier than ceramic biomaterial and elasticity module is 45 GPa which has at near elasticity module of human bone [10]. Magnesium - aluminium alloy has been used as biomaterial, for example AZ31 (Mg-3Al-1Zn), contains of Zn as additional material [11]. The specimens hold in VMC machine and cut into dimension of 40 x 50 x100 mm. First full drilled with twist drill 5 mm diameter and the depth of 40 mm for pre drill with 405 rpm and feeding 0.1 mm/rev. in order to get straight path while drilling of diameter 12 mm. A drill of 12 mm diameter with parameter and different level, inserted so that drilling process happened into each specimen holes measurement and follow the rule of Taguchi L18.

The surface roughness every hole must be checked by CMM or with USB microscope so that roughness grades each surface will be known. Cylindricity and perpendicular surface of every holes checked by CMM so that the roundness or the perpendicular will be known. Surface roughness each hole has to be examined with CMM or USB microscope, and the rate of holes roundness and its perpendicular can be determined too. In general, the surface roughness value resulting in range 0.33 until 1.99 μm. The highest surface roughness value 1.99 μm obtained in the drilling process in 405 rpm, feeding 0.1 mm/rev, point angle 450 and synthetic oil lubricant. The highest surface roughness value is 1.99 μm obtained in drilling process happened in rotation 405 rpm, feeding 0.1 mm/rev, point angle 450 and synthetic oil lubricant. While the lowest surface roughness value obtained in 0.33 μm happened in rotation 890 rev/men, feeding 0.2 mm, point angle 650 and synthetic oil lubricant.

The holes cylindricity value produced be range of 0.017 mm to 1.122 mm. The value of cylindricity has the highest poverty rate 0,017 mm where is obtain on drilling process n = 405 rev/men, feeding of 0.3 mm/rev and point angle of 450 with lubricant of crude palm oil. While the lowest value of cylindricity obtained in 1,122 mm in drilling process at spindle speed of 405 rpm, feeding of 0.3 mm/rev and point angle of 550 with soybean oil lubricant. The value of perpendicular is in (run no 14) that was between rotation 890 rpm, feeding of 0.2 mm/rev, point angle of 550 and lubricant with soybean oil. While the farthest deviation to the process is number 2 where happened in rotation of 405 rpm, feeding of 0.1 mm/rev, point angle of 550 and lubricant is palm oil.

## 3. Results and Discussions

### *3.1. Surface roughness value of drilling process*

In principle the surface roughness objects work on the machine not uniform throughout the machining process because affected by the state of cut as, the rate cuts, depth of cut in cutting and types of cutting tool.

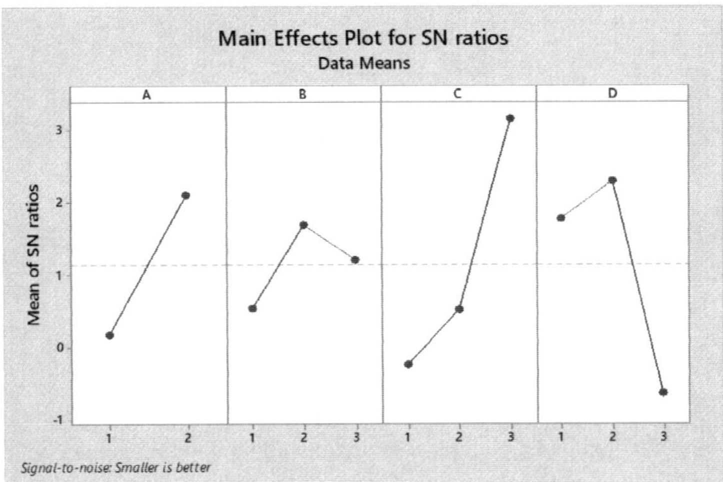


**Figure 1.** SNR Chart plot for the priority factors of surface roughness (Ra) for Magnesium AZ31.

Figure 1 that point angle (delta 3.4023, rank=1) has leverage greatest in SN Ratio, followed by a lubricant (delta 2.9504, rank 2) followed by round and feeding. The value of the roughness. The surface roughness value not only influenced by a feeding factor and spindle speed. It's also influenced by the tool nose radius factor [12]. Based on formula Ra = f / Re, so the greater the tool nose radius, the surface roughness will be reduced. But a rough surface is because the edge wears contours uneven/smooth. Mechanically there are parameters that can affect Ra namely feeding (f) and nose radius (Re), if f increased and Re move down there will be increasing Ra.

### *3.2. Cylindricity value of the drilling operation*

One of the constituents of a change in roundness value would be happened if the length of left cutting edge and right cutting edge are not the same. On the lowest roundness value obtained to the process with the test number 8 with the 1.122 mm in drilling process with spindle speed of (n) 405 rpm, feeding of 0.3 mm/rev and point angle of 550 and the lubricant is soya oil. As shown in Figure 2 that a drill bit for test number 8, on its cutting edge, one side on the size 12.95 mm and another side is 13.25 mm happened the difference between long body so that these at the time spun and cut will cause tangential force who not in equipoise so that a circle being oval do not a circle geometrically. The difference between cutting force resulting to an insistent wall a hole in line with a style of cut into the larger [13]. The point angle (delta 3.4023, rank=1) has greatest leverage in SN ratio, followed by a lubricant (2.9504, rank=2) followed by spindle rotation and feeding, as shown in Figure 3.

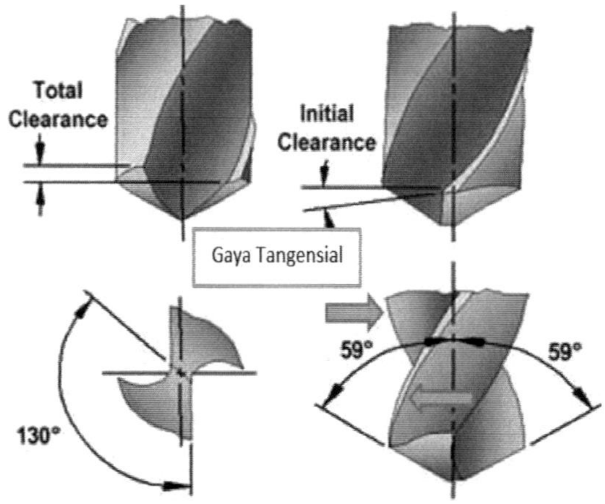


**Figure 2.** Tangential force on drill bit

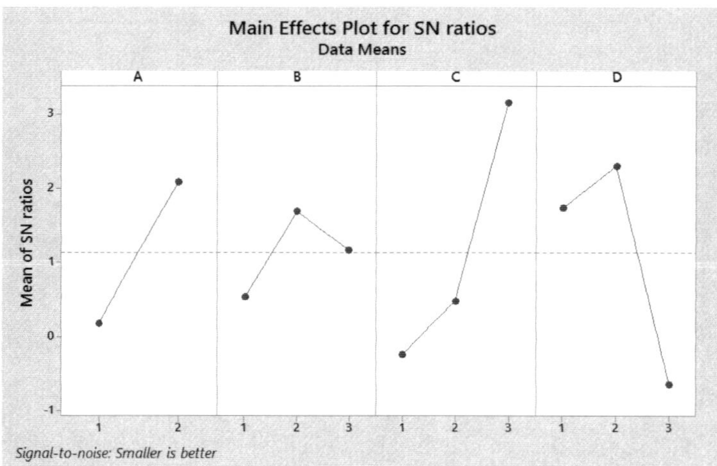


**Figure 3.** The plot SNR ratio to the cylindrical main factor

### *3.3. Perpendicularity of the drilling operation*

The result of optimizing the ratio SN ANOVA found a significant parameter against fixed variable of perpendicular seen of the value of P-value < α (0.05) used in research. In this case feeding 1 with p = 0.044, and interaction between (feeding * lubricant) with P=0.041 which is smaller than α (0.05) showed the significance. This showed that feeding 0.1 mm/rev to the drill machining process of magnesium AZ31 use a twist drill diameter 12 mm influence on the result perpendicular value from drilling process, especially if there is the interaction of the feeding of 0.1 mm/rev and synthetic lubrication. Viscosity influence at the speed of a liquid filling contact between machine tool and a work piece and a liquid thin of them [14]. Hopefully with a feeding of 0.1 mm/rev (smallest one) with synthetic lubricant and with high viscosity caused more stability of cutting holes wall that led to good perpendicularity value as shown in Figure 4.

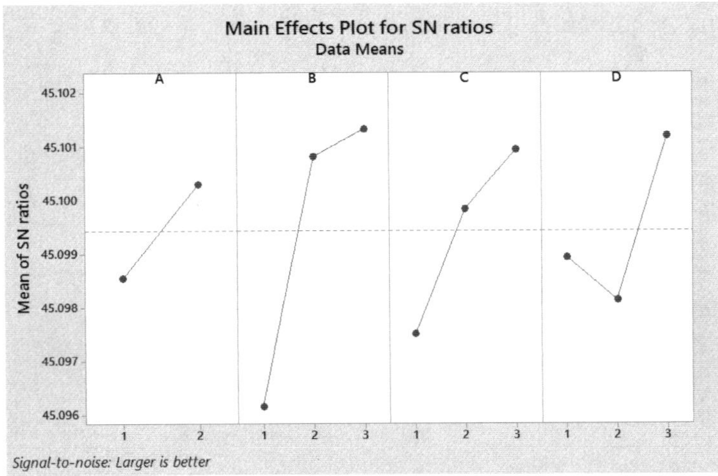


**Figure 4.** The plot SNR ratio to the perpendicularity main

Feeding (delta 0.01, rank=1) have greatest leverage in the SN Ratio, followed by point angle (delta 0.00, rank=2) are followed by lubricant and spindle speed. If we use feeding 1, lubricant 1 associated with the highest mean strength. The result of two directions ANOVA shows that interaction between feeding and lubricant is significant. Perpendicular deal with the machine tool structure with tool geometric (length and helical corner)

## 4. Conclusions

The research results show that surface roughness of machined surface affected by contribution of point angle of 18.9% and lubricant of 14.5%, where the point angle is 650 with the synthetic oil lubricant. The emergence of a tendency nose radius resulting in increasing subtlety level. Roundness influenced by point angle 450 of drilling process, especially if there is an interaction to feeding 0.2 mm/rev and synthetic lubricant. The accuracy of cutting edge led to tangential force so that had an impact on Cylindricity their rotation stability and cylindricity drilling result. The perpendicular value of magnesium AZ31 is strongly influenced by cutting parameter of feeding that produces significance p=0.044, while strongest significance happened if there were interaction between feeding 0.1 mm/rev and synthetic lubricant in p=0.0041. The less feeding of the process with a high viscosity of synthetic lubricant cause more stable of hole wall cutting and led to good value of perpendicular

## References

[1] Chong K Z, Shih T S 2002 Optimizing Drilling Condition for AZ61A Magnesium Alloy *J. Material Transaction* **43** (8), pp 2148-2156
[2] Chen and Fuh 1996 *J of Material and Process Technology* Vol pp 314-322.
[3] Chen and Tsao 1999 *J of Material and Process Technology* **88** pp 203-207
[4] Haan and Batzer 1997 *J of Material and Process Technology* **71** pp 305-313
[5] Wiriyacosol, Armarego 1979 Annual CIRP **28** pp 87-91
[6] Biermann D, Liu Y, 2014 Innovation Flow Drilling on Manufacture of light weight Components El Sevier Academic Process Series Procedia CIRP **18** pp 209–214

[7] Hofmann D 2009 Knowledge-Based Approach toward Hydrolytic Degradation of Polymer-Based Biomaterial WILEY-VCH Verlag GmbH & Co. KGA Weinheim
[8] Spicer 2014 Tool Wear Monitoring for Ultrasonic Metal Welding of Lithium-Ion batteries Manufacturing Reasearch Laboratory Michigan
[9] Ibrahim G A, Arinal H, Jamiatul A 2016 Wear of carbide insert at machining of Inconel 718 using Minimum Quantity Lubrication Proc. Nat. Seminar of Mechanical Engineering XV Bandung
[10] Witte F, Hort N, Vogt C, Cohen S, Kainer K U, Willumeit R I 2008 Degradable Biomaterials Based on Magnesium Corrosion Current Opinion Solid State Materials Science **2** pp 63-72
[11] Kirkland N T, Lespagnol J, Birbilis N, Staiger M P 2010 A Survey of Bio-Corrosion Rates of Magnesium Alloys Corrosion Science **10-52** pp 287-291
[12] Ibrahim G A 2014 Identification of surface roughness value at machining of magnesium alloys *J. of Mechanical* **5** no 1 pp 11
[13] Koenisberger 1964 Design Principles of Drozka Machine tools Pegamon Press Oxford
[14] Anton 2018 Viscosity of metal Working Fluids in Cutting Processes SVMTM Series (Anton-Paar GMBH)

IOP Conf. Series: Materials Science and Engineering **857** (2020) 012015 doi:10.1088/1757-899X/857/1/012015

# Identification and plastic type and classification of PET, HDPE, and PP using RGB method

**I Yani[1*], D Rosiliani[1], B Khona'ah[1] and F A Almahdini[1]**

[1] Department of Mechanical Engineering, Sriwijaya University, Palembang, South Sumatera, Indonesia

***Email:** irsyadiyani@ft.unsri.ac.id

**Abstract**. Plastic is one of the most commonly used materials in daily life. The characteristics of the material that is light, durable, waterproof, and cheap become the main reason why plastic is chosen in the manufacture of ready to use goods. However, due to the increasing use of plastic materials, the amount of waste produced is significant, and tends to increase over time. Therefore, an effective process and recycling of plastic waste are needed. This paper will discuss a method of identifying and classifying plastic waste by type, that use three types of plastic, there are PET, HDPE and PP. The method used in this study is the RGB method, that based on the color value of the plastic.

**Keywords:** RGB, RGB method, identification, classification, plastic, PET, HDPE, PP

## 1. Introduction

Plastic is the material that we can easily find at our daily life because of its characteristics. In some uses, plastic is only used at once, such as plastic bottles. Moreover, the most used plastic bottles are PET (Polyethylene terephthalate) types [1]. The continued use is increasing that make the amount of plastic waste continue to increase over time. Plastics require at least decades, even some types of plastic need hundreds of years to decompose.

The accumulation of plastic waste that occurs continuously, if it is not balanced with proper processing will cause pollution. This pollution will be very detrimental because it can affect every aspect of living thing. Moreover, it can contaminate the soil, water, and other elements of life.

Recycling is an effective way to process plastic waste into something kind of useful. Plastic mixes are one of the materials that are difficult to recycle due to their complex characteristics [Wang, 2015]. Until now, there are at least two methods in sorting, namely the manual method and the automatic sorting method. Sorting plastic waste is based on the type of resin forming it. It is essential because the sorting process must be obtained by pure resin from each type of plastic.

The types of plastics consist of 7 types; there are polyethylene terephthalate (PET), high-density polyethylene (HDPE), polyvinyl chloride (PVC), low-density polyethylene (LDPE), polypropylene (PP) and polystyrene (PS), marked with numbers 1 to 6 in each type. Also, there is one additional type of resin to indicate plastic other than the previous six, marked with code seven called "other" [2]. If two or several types of plastic resin are mixed, it will cause a negative impact that can harm the environment, as if PVC and PET are mixed up, it will produce hydrochloric gas [3, 4].

Plastic waste is classified based on the resin or its constituent material. However, nowadays, the manual system is barely used and start to change to the automatic systems that are more efficient and easier. In general, computer vision aims to take pictures, standard lighting systems, and as software for image processing and analysis using digital cameras [5]. Some studies that use computer vision as a technology for assessing the quality of a product, such as Benalia et al. [6]; Jackman et al. [7]; Ercisli et al. [8]. These studies show that computer vision is an accurate technology, not damaging, consistent, easy to use, cheap, and efficient.

## 2. Methodology

In 2019, Bekti Khona'ah et. all. have researched the identification and classification of plastic types using the RGB method. However, in her research only two types of plastics were used, there are PET and HDPE [9]. While in this study uses three types of plastic so that the results of this research can be more accurate. The first thing to do is take the image of plastic using a 720 mega pixel webcam. The webcam is installed at a distance of 600 mm from the plastic. The image was taken through a 640 x 480 pixels webcam because it is considered to be most suitable for the needs in terms of image color retrieval. The plastic image that has been detected by the webcam will be directly sent to the PC that is connected to the webcam. The format used to store images of plastic is jpg or jpeg format. This format is chosen because jpg or jpeg is a bitmap file compression scheme, that can compress large sized images into smaller sized images. It is advantageous because it can compress extensive data in rather small storage media.

After the required image of plastic is complete, it can be used to create a database. The making process of the database is based on the color image using the RGB method, that was previously carried out by the characteristic feature extraction process from the plastic. RGB is a method consisting of three primary colors or basic colors, namely red (R), green (G) and blue (B). The color value of each primary color in this RGB method is 0 to 225, so the number of colors obtained is 256 x 256 x256 or about 16.7 million types of colors. RGB is illustrated using a three-dimensional cube based on the Cartesian coordinate system, as shown in Figure 1. Where specific colors will be obtained if we combine the primary colors that exist (fig 2) [9]. Processing of the RGB method is complete when we already obtain the values from the three primary colors that are conditioned in a specific pixel.

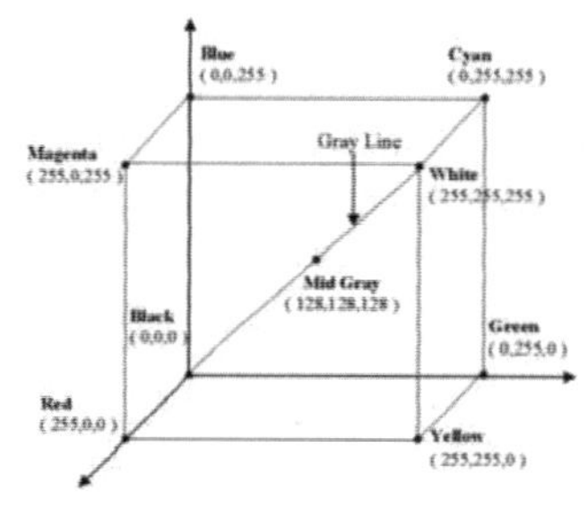


**Figure 1.** RGB illustrated based on Cartesian coordinate

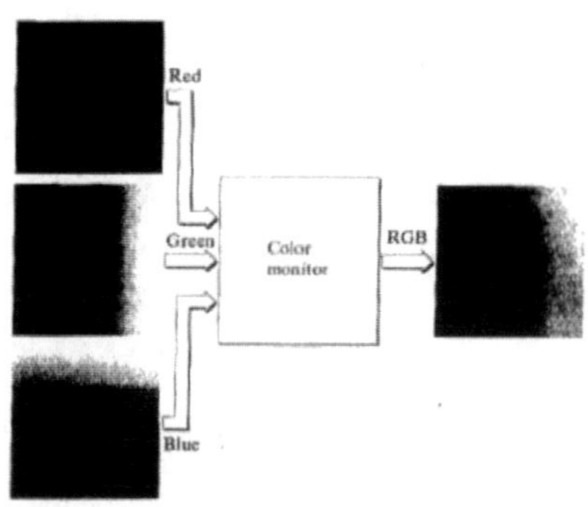


**Figure 2.** Colors input in RGB method

Preparation process of the database is based on the characteristics of the primary colors red, green, and blue in the form of universal values are of each color in each plastic image. The average of each color red, green and blue can be obtained by the average formula, for red is the total of all red values is divided by the amount of data available, as well as the green color and the blue color.

The first step in making a data base is to set the template using the im crop command with settings [280 180 33 33], and the image of plastic will be the size of 5x 5 pixels, and we get what we want. Image feature extraction is a fundamental process in image identification and classification. Only certain features needed are maintained for the recognition process so that classifications can be implemented under the conditions required. Feature extraction is a way to simplify the for and make it easier for the system to learn the target concept, and get a simpler hypothesis. The choice of feature extraction method is the essential factor in carrying out a pattern recognition system.

Several factors, such as lighting, camera position, treatment (contrast and brightness), camera height above the object, will affect the quality and sharpness of the final image we obtained, so that configuration is needed in such a way as to get the best results.

After the process of the main database, making has been completed; the plastic identification testing process is carried out by taking a webcam image. The results of the identification are in the form of graphs of red, green, and blue values, that will be processed for the classification process of the type of plastic.

## 3. Result and Discussion

### *3.1. System Description*

The system of classification and identification of automatic plastic sorting systems consists of three main components; there are a computer system, a webcam, and an integrated automatic system. The description of the system is a description of the design that has been integrated with the automatic identification and classification system into mechanical systems. The webcam is used in the process of taking images from the objects in the form of several types of plastic; then the image will be processed into digital format. Digital images that have been obtained from the image retrieval process will be processed by a computer system that aims to obtain data similarity values with the database that has been made before. The next process is the automatic identification and classification of computer systems according to the type of plastic based on color images.

### *3.2 Data Base*

Database making is one of the most important processes because it will be used as a basic framework in developing automatic sorting systems. The first step is to create an object database. In making a data base, the process that must be done consists of image pre-processing, feature extraction, and classification of color images in the image of plastic that has been obtained before. Image pre-processing consists of image capture and image acquisition processes, and feature extraction consists of cropping and extraction of images, and the trimming process consists of matching templates. The extraction process includes the edge extraction process and the color extraction process. The results of several classifications show the harmony between features and values in the data base.

### 3.3 *Data Pre-Processing*

The first step is the process of image taking. The quality of the image affects the quality of raw data, and if raw data has good quality, it will automatically affect the overall system performance. Image capture uses a digital webcam as a sensing device, where the digital webcam can produce images with high quality and low noise.

The image is adjusted to 24-bit color, and 640 x 480 resolution obtained using a webcam, then stored in the jpg format. After the image is obtained through the image retrieval process, the quality of the image is examined, whether it is good enough or not. If the image quality is not good enough, the image cannot be processed. As a result, the existing data base is only filled with templates with good quality, and the accuracy of the system will also be better. All images that used are good quality images so they can be processed entirely. It is because at the time of the image taking process, the researchers positioned the plastic to a predetermined position.

**Figure 3.** Object positioning

### *3.4. Image Extraction*

Image extraction is included trimming parameters and pruning solutions needed in the image composition calculation process. Image solution space still has many parts that must be eliminated because it includes some part of its background that can be identified and decreasing the accuracy of image values. Removal of unnecessary parts of the image will simplify the identification process because it does not require the calculation of the composition features, and the image area exception feature only needs to be counted once for images. The research template settings that used is im crop [280 180 33 33] so that the final size of the image becomes 5 x 5 pixel. Examples of images after the extraction process can be seen in the picture.

**Figure 4.** Cropping image use im crop

### *3.5 RGB Value*

After the identification process is obtained, the values of each color are red, green, and blue for each type of plastic. Moreover, it can be seen in the following pictures (figures 5, 6, and 7).

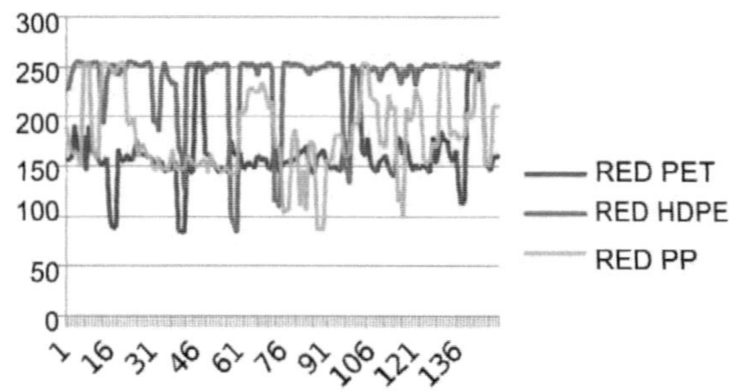


**Figure 5.** Graph value of red on PET type and HDPE type with RGB method

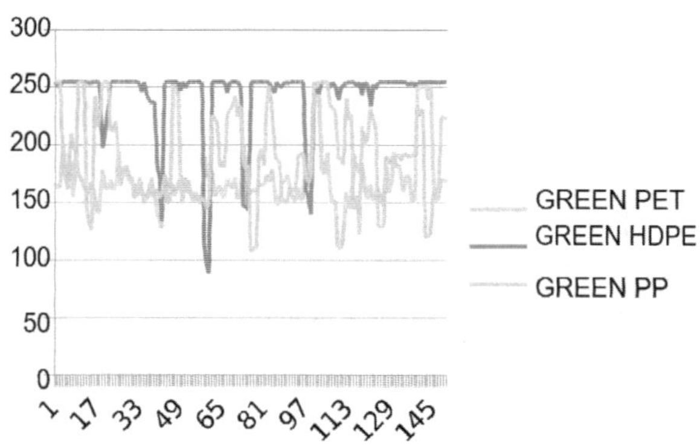


**Figure 6.** Graph value of green on PET type and HDPE type with RGB method

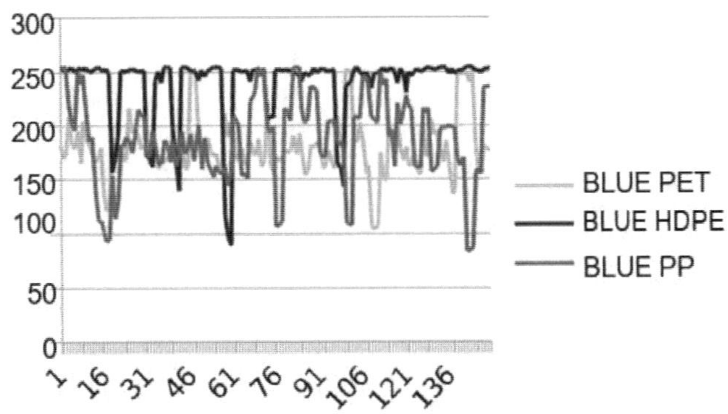


**Figure 7.** Graph value of blue on PET type and HDPE type with RGB method

Based on the profile on Fig. 5-7 above, PET type plastic has a maximum red color value of 253.2042 and minimum value of red 145.39, while the maximum color value for green is 254.9602 and the minimum color value for green is 151.7621, then for maximum color value blue is 252.4066 and the minimum color value for blue is 160.128. Moreover, for HDPE type plastic has a maximum color value for red ranging from 255 and a minimum color value of red 132.8685, for maximum green 255 color value and minimum green color value 140.109, and the maximum blue color value is 254.7362, and minimum color value for blue is 144.1185. The maximum value of red for PP type is 254.1367 and the minimum value is 85.9974, the maximum for green is 254.8088 and for the minimum is 149.3564, moreover for the blue value the maximum value is 236.3651 and for the minimum value of blue is 107.455.

It can be seen that the comparison of the value of red, green, and blue of PET, HDPE and PP seems quite clear, plastic with HDPE type has a value of red, green, and blue that is larger than the type of PET and PP. Based on the results obtained, the difference in the value of three types is significant; the possibility to distinguish each type of plastic will be more significant and more accurate.

The average color value of the PET type is as follows, for the average red value of 173,758, the average green value is 182,247, and the average color value for blue is 191,612. Meanwhile for HDPE type plastic, the average color value for red is equal to 232.44, the average color value for green is 236., and the average color value for blue is 233.45. And the average color of PP type is 172.39 for red, 197.28 for green anda 185.418 for blue.

PET type plastic, HDPE type plastic and PP type plastic have a different average value for each color. Comparison of the average value of PET and HDPE is 58,682 for red, 53,853 for green and 41,838 for blue. Comparison of the average PET and PP 1,368 for red, 15,034 for and 6,192 for blue. Comparison of the average value of HDPE and PP is 60,01 for red, 38,82 for green, and 48,032 for blue. Based on the comparison of the three types of plastic, it can be seen that there are significant differences between the average color values of three of it.

## 4. Conclusion

The type of plastic used in this study includes three types of plastic, the type of PET, HDPE andThe study was conducted to present the process of identifying and classifying three types of plastic. Identification and classification are carried out using the RGB method with a data base obtained from image pre-processing, feature extraction, and color image classification in the image of plastic. Set the template by using the im crop command with settings [280 180 33 33], until having the plastic image with the size of 5 x 5 pixel according to what you want. Red, green, and blue values have comparisons that appear to be quite clear between three types of plastic. Plastic with HDPE type has a value of red, green, and blue, that are bigger than the PET and PP type. Comparison of the average value of PET and HDPE is 58,682 for red, 53,853 for green and 41,838 for blue. Comparison of the average PET and PP 1,368 for red, 15,034 for and 6,192 for blue. Comparison of the average value of HDPE and PP is 60,01 for red, 38,82 for green, and 48,032 for blue. Based on the comparison of the three types of plastic, it can be seen that there are significant differences between the average color values of three of it.

**References**

[1] W. Srigul, M. Kupimai, and P. Inrawong, "Plastic classification base on the correlation of RGB color," 2016 13th Int. Conf. on Electrical Engineering/Electronics, Computer, Telecommunications and Information Technology (ECTI-CON).
[2] C. Chow, W. M. W. So, and Tsz-Yan Cheung, "Research and development of a new waste collection bin to facilitate education in plastic recycling," Journal Applied Environmental Education
[3] D. A. Wahab, A. Hussain, E. Scavino, M.M. Mustafa and H. Basri, "Development of a Prototype Automated Sorting System for Plastic Recycling," American Journal of Applied Sciences **3** (7): 1924-1928, 2006
[4] B. Ruj, V. Pandey, P. Jash, and V. K. Srivastava, "Sorting of plastic waste for effective recycling," Int. Journal of Applied Sciences and Engineering Research, **4**, Issue 4, 2015
[5] E. M. Oliveira, D. S. Leme, B. H. G. Barbosa, M. P. Rodartec, and R. G. F. A. Pereira, "A computer vision system for coffee beans classification based on computational intelligence techniques," Journal of Food Engineering **171**, February 2016, Pages 22-27
[6] S. Benalia et al, "Computer vision for automatic quality inspection of dried figs (Ficus carica L.) in real-time," Computers and Electronics in Agriculture **120**, January 2016, Pages 17-25
[7] P. Jackman, D. W. Sun, and P. Allen, "Recent advances in the use of computer vision technology in the quality assessment of fresh meats," Trends in Food Science & Technology **22**, Issue 4, April 2011, Pages 185- 197S.
[8] Ercisli, B. Sayinci, M. Kara, C. Yildiz, and I. Ozturk, "Determination of size and shape features of walnut (Juglans regia L.) cultivars using image processing," Scientia Horticulturae **133**, 6 January 2012, Pages 47-55
[9] Khona'ah, B., Rosiliani, D., and Yani, I., 2015. Identification and Clasification of Plastic Color Images Based on The RGB Method.

# Temperature distribution simulation on Aluminum incineration furnace using Autodesk simulation mechanical CFD 2018

**I Thamrin[1,*], I Yani[1], F H Yusuf[1], Z Abidin[1], T Maseko[1]**

[1] Department of Mechanical Engineering, Engineering Faculty, Universitas Sriwijaya, South Suamatera, Indonesia

E-mail: ismailthamrin@ft.unsri.ac.id

**Abstract.** Used lubricating oil is one of the liquid wastes which is produced by an industrial machine or private vehicle machine. Lubricating oil is a waste that is produced by machines in both large industrial and private vehicle engines. This research aims to examine the process of manufacture and the system of oil-fueled crucible stoves, and simulated the temperature at aluminium melting furnace by doing some form of design testing; selection of furnace material using carbon steel, brick, plate; furnace modeling which has three parts of crucible, fireproof stone and lids where the three parts have different sizes and uses of each, then set the simulation as what we aim to such as the temperature distribution where the air temperature reached 40°C heat source from combustion 1200°C. From the combustion, the velocity value obtained is 0.5 in/s, distribution at crucible is 749°C. In conclusion, the used oil smelting furnace can melt used aluminium metal, so that the used metal oil melting furnace can be used in the home industry to recycle used aluminium metal.

**Keywords**: Temperature distribution, CFD, furnace, alumunium

## 1. Introduction

The development of the smelting and metal casting industry in Indonesia is currently very low. Even though Indonesia has the potential to become one of the largest markets in the world. Many small-scale metal casting industries are constrained by their development, this is due to the fact that metal smelting kitchens that are available on the market are very expensive and difficult to obtain because they have to be imported from abroad. The choice of the type of kitchen and the smelting process used must be in accordance with the type of metal chosen and also according to the desired product. The types and classifications of smelting kitchens that are currently developing include Crucible kitchens, cupola kitchens, electric arc kitchens, induction kitchens, converter kitchens, and the Thomas and Bessemer kitchens. The fuels used are also diverse including coal, fuel oil, electricity, charcoal, even gas-shaped fuels [1].

To solve the problems, it is necessary to design and manufacture metal smelting kitchens that are simple, easy to make, easy to move (portable) and the most important point, is that the machine has acceptable price so that they can be reached by household scale casting industries. The metal smelting

IOP Conf. Series: Materials Science and Engineering **857** (2020) 012016 doi:10.1088/1757-899X/857/1/012016

kitchen resulting from the design and manufacture will have several advantages including fuel efficiency because it uses used oil fuel and the combustion chamber temperature reaches 1000°C because kitchen construction uses heat insulation of refractory bricks. For this reason, it is necessary to obtain a relatively inexpensive aluminium smelting kitchen design using used oil fuel. Aluminium has a melting point of 660°C , so with a kitchen temperature of 1000°C it is very potential to be used to melt aluminium. Other types of metals that can be melted with melting points below 1000°C are copper (962°C), brass (940°C), tin (327°C), magnesium-lead (630°C), Sn-Bi (232°C), etc. [2,3].

In waiting for the smelting, the temperature produced must be greater than the melting point of aluminium, which is 660°C. In this research the temperature caused by heat loads on the furnace will be simulated using computation fluid dynamic (CFD) software.

Stoves or hearth are tools used to melt meals to smelting or heated materials to change the form (example: rolling) or the properties (heat treatment) [4].

Minimum-scale and Medium-scale alumminium melting and usually done by crucible furnace. A characteristic of crucible furnace is the usage of vessel to place the metals that are going to be melted. The vessel has a form of pot-like ceramic crucible with upper larger diameter which acknowledged by the name of kowi. This furnace is differentiated by the fuel-type that are being used, such as charcoal, oil, and gasses. Another differentiation is being done according to the furnace;s construction, which are divided into furnace with non-fixed kowi, and furnace with fixed-kowi and tungkik furnace [1].

In the melting furnace with the direct fire combustion, the fire flash is directly touching the alumminium scrape crucible. The study of the distribution of temperature in the melting furnace is very important in order to know the location of stresses and strain caused by increasing the temperature in the furnace.

## 2. Method

Furnaces are devices used for high temperature heating. Heat energy for furnace fuel can be supplied directly by burning fuel, by electricity such as electric arc furnaces, or by induction heating in an induction furnace.

Heat treatment furnaces are heating stoves used to heat the metal until it reaches its melting point so that it can be used for casting. Heat energy that is needed are including: heating energy to raise the temperature to the melting point and fusion heat to convert metals from solid to liquid.

Based on the usefulness of the heat treatment furnace divided into four (4) namely, annealing furnace, hardening furnace, tempering furnace, and carburizing furnace. The difference from the four heat treatment stoves this is the capacity to achieve its temperature. According to heat sources, the heat treatment furnace can be divided into two: fuel burning furnaces fuel-fired furnaces (see Fig. 1 and 2) and electric heating furnace electrical heated furnaces. Fuel combustion stoves can be further classified depending on the type of fuel, solid fuel (figure 3), liquid fuel, and gas fuel. Liquid fuel commonly used is fuel oil. As it concerns on efficiency and economy, gas combustion stoves are lower than on electric heating stoves [5].

Electric furnaces have advantages such as uniformity of temperature in the chamber, free of pollution, neat and clean working conditions, efficient use of heat energy (minimizing lost energy), and easy start and closing down. Electric furnaces can be grouped into resistance furnaces, arc furnaces, Induction furnaces, plasma arc furnaces, and electron beam furnaces [6].

Heat is a form of energy that can be transferred from one system to another due to temperature differences. Heat Transfer always from media with higher temperatures to lower temperatures, and heat transfer stops when the two media reach the same temperature. Problems with heat transfer those encountered in practice can be considered in two groups: (1) assessment and (2) size problems. The assessment problem is related to determining the heat transfer rate for a system that is at a certain temperature difference. The problem of size relates to determining the size of the system to transfer heat at a level determined for a certain temperature difference.

IOP Conf. Series: Materials Science and Engineering **857** (2020) 012016 doi:10.1088/1757-899X/857/1/012016

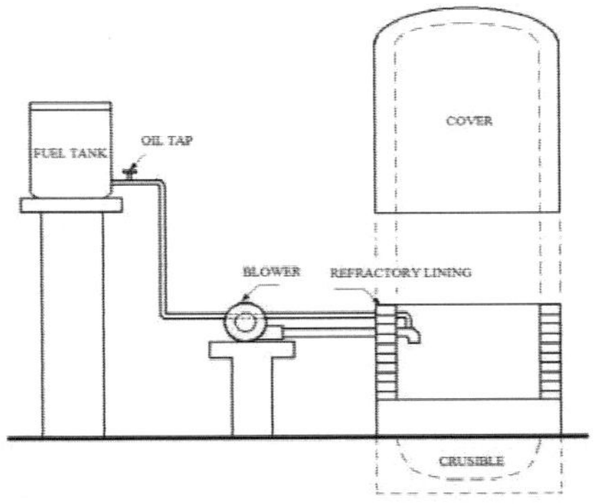


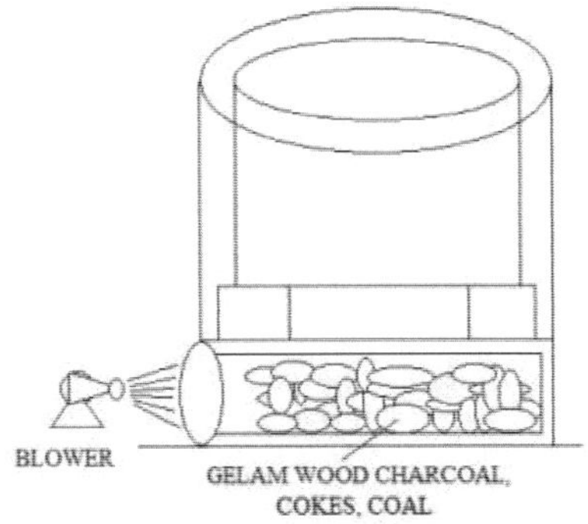


**Figure 1.** Incinerating Furnace and Fuels **Figure 2.** Furnaces and Solid Fuels

Conduction heat transfer on a flat plate only at the X-coordinate is shown in Figure 6. Based on Fourier's law equation about conduction is as follows [4,6].

$$q_x = -K_{xx}\frac{dT}{dx} \tag{1}$$

In which:

$Q_x$ = X-axis transfer heat
$K_{XX}$ = Thermal conductivity on the X-axiskW(m.°C) atau Btu/(h-ft-°F)
dT/dx = temperature gradient°C/m

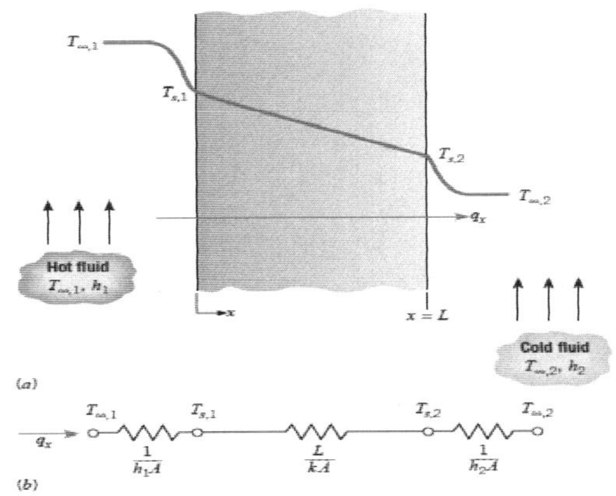


**Figure 3.** Heat Transfer at Flat plate

The process of heat transfer in the furnace will begin at the heat transfer from the fire which is sprayed directly through the furnace lid hole to aluminum on the base of the crucible which is the transfer of heat of convection.Conduction heat transfer occurs between krusibel used aluminum, also in the crucible with 0.3 m thick conduction heat transfer occurs.

## 3. Result and Discussion

After the run simulation process is complete, the temperature to time graph will be obtained from two simulated crucible materials. The CFD simulation on the furnace was successfully carried out using Autodesk CFD software . The speed vector in the dimensional field view is shown in figure 4. In the figure it can be observed that the maximum velocity distribution of high fluid . If viewed from thermal conductivity the Graphite material has a value high. The maximum speed is shown in the flow of flame when entering the combustion chamber, after the combustion energy is released then the speed increases and continues to decrease towards the outlet.

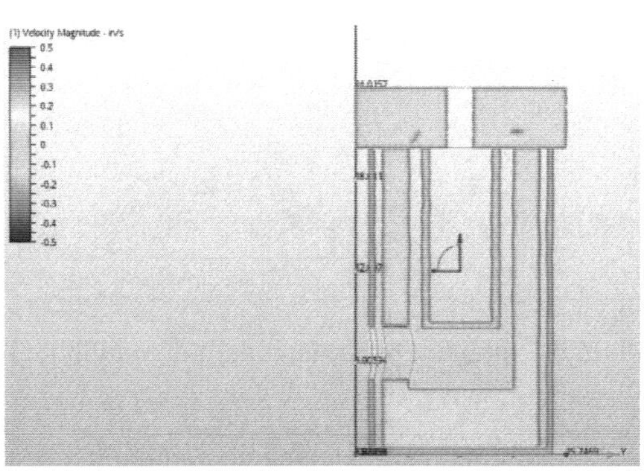

**Figure 4.** Rate at graphite kowi material

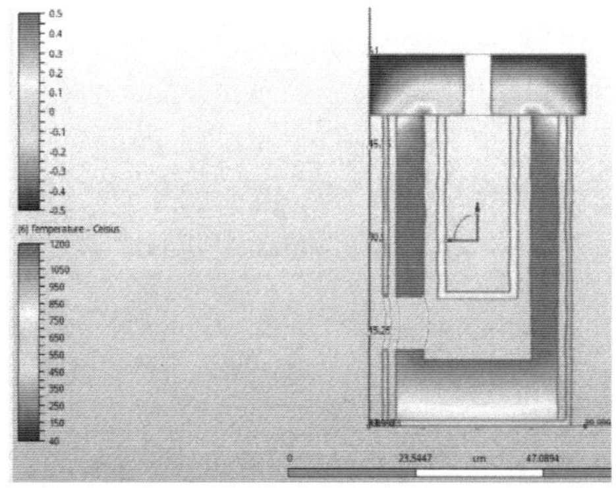


**Figure 5.** Termperature distribution

Figure 5 shown the results of the simulation of the temperature distribution in each material. Visible range minimum and maximum temperature ranging between 0°-1200°C.

In figure 6, the following shows the temperature distribution from the inlet to the outlet in the kowi material Graphite maximum temperature is found in the inlet furnace, the temperature decreases more drastically when heading to the outlet, which is close to the value of 748°C.

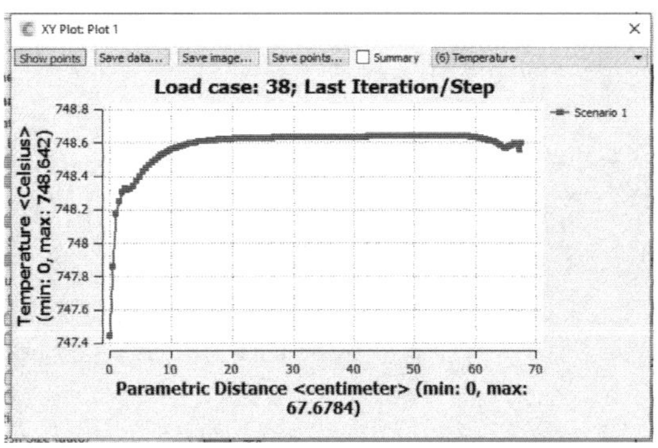


**Figure 6**. Temperature distribution graphic from inlet to outlet at graphite kowi material

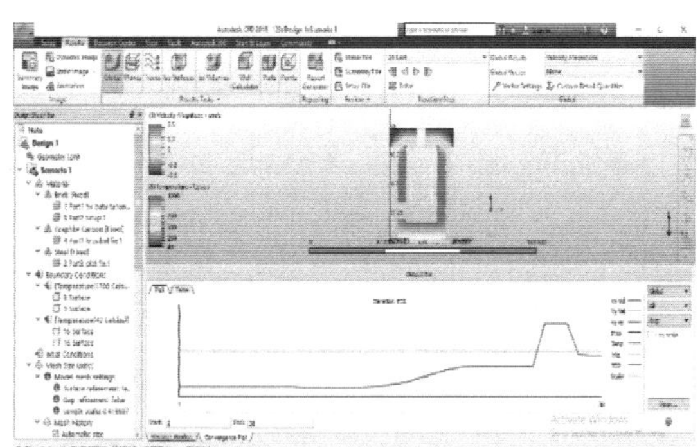

**Figure 7.** Combustion on steel material

The results of the temperature distribution that has been simulated indicate an increase in temperature during the combustion time (figure 7). Furnace is the increase in the maximum value and minimun value with increasing time, as highlighted by others [6,7].

Based on the simulation results it can be seen that the crucible material affects the graph of temperature distribution in the aluminum melting furnace. In the results of simulations that have been done, it is found that crucible with steel materialcarbon can increase the temperature with a maximum value of 1200$^0$C and a minimum value of 40$^0$C.

## 4. Conclusion

Based on the conclusions obtained from the research and discussion on the Simulation of Temperature Distribution in Aluminum Melting Furnaces by Using Autodesk Software CFD 2018, are as follows:

Based on the results the combustion temperature of used oil is quite high (the highest temperature measured is $1200^0$ C. With kowi construction using used oil carbon steel material can be used as fuel to melt aluminum in aluminum melting furnaces. Based on the simulation results, it is found that crucible with steel material carbon undergoes a change in temperature distribution from transient to reach steady state quickly in aluminum melting furnaces.

## References

[1] Joseph, O.,& Irabodemeh, J. M. (2016). *Design and Thermal Analysis of Crucible Furnace For Non –ferrous Metal,* **6**(3), 5-9
[2] [3] Callister Jr., W.D., 2007. Materials Science and Engineering, 7th ed. (New York.: John Wiley & Sons Inc)
[3] Totten, G. E. & MacKenzie, D. S. 2003. *Handbook of Alumunium Vol.1 Physical Metallurgy And Proceses*. (New York: Marcel Dekker, Inc)
[4] ASTM G1-03, 1999. Standard Practice for Preparing , Cleaning , and Evaluating Corrosion Test, in: ASTM 03.02 Wear and Erosion Metal Corrosion. ASTM International, Pennsylvania, pp. 1–9
[5] Adetunj, O. R., Kuye, S. I., & Also, M.J. 2015. *Microstructures of Mild Steel Spring after Heat*, (Desember 2013).
[6] Logan, D. L. 2017. A First Course in the Finite Element Method, 4th ed. Chris Carson.
[7] Holman, J. P. 2010. *Heat Transfer,* (New York: The McGraw-Hill Companies, Inc.).

# The analysis of straw mushroom potential development using an empty fruit bunches materials

**Sarono[1*], Yana Sukaryana[2], Zainal Arifin[1], and Sri Astuti[3]**

[1]Lecturer of Agricultural Technology Department in Politeknik Negeri Lampung, Jalan Soekarno Hatta No. 10 Rajabasa Bandar Lampung 35144, Indonesia

[2]Lecturer of Animal Husbandry Department in Politeknik Negeri Lampung, Jalan Soekarno Hatta No. 10 Rajabasa Bandar Lampung 35144, Indonesia

[3]Lecturer of Economics and Business Department in Politeknik Negeri Lampung, Jalan Soekarno Hatta No. 10 Rajabasa Bandar Lampung 35144, Indonesia

*Email: sarono@polinela.ac.id

**Abstract.** The empty fruit bunches (EFB) known as a waste material from crude palm oil (CPO) production. It has widely studied for the potential used of EFB. One of products produced from EFB with development potentials is straw mushroom. The objective of this research was to analyze the potentials of EFB material and straw mushroom based on EFB material in Lampung province. The result showed that the EFB potential in Lampung province was 111,144 ton annually and straw mushroom was 4,835 ton annually. The biology efficiency ratio of EFB into straw mushroom in production scale was averagely 3.93%. Districts having potentials for developing straw mushroom business by using EFB material were Mesuji, Middle Lampung, Tulang Bawang, Way Kanan, and North Lampung.

**Keywords:** straw mushroom, EFB, and biology efficiency ratio

## 1. Introduction

Straw mushroom (*Volvaria volvaceae*) in Indonesia is a very popular agricultural product and its demand is increasing [1]. Many benefits of straw mushroom when it is added into meals include increasing amino content, increasing volatile component contents, reduced peroxide value, and improving food taste [2].

Straw mushroom cultivation is a most efficient and economical biotechnology process in changing agroindustry material with lignocellulose content into nutritional and functional food [3]. Many agricultural by products, that are left unusable and they even damage environment, can be used as planting media for straw mushroom.

One of by products in the conversion process of fresh oil palm bunches into crude palm oil (CPO), that is not yet optimally used, is empty fruit bunches (EFB). According to [4], about 16% - 23% of EFB is produced by fresh oil palm bunches (FFB). It means that 1 ton of processed FFB shall produce 160 kg to 230 kg EFB. [5] Suggest that the average of produced EFB is 23% from total of processed FFB.

The main compositions of EFB are 14.1%-30.45% lignin, 23.70%-65.0% cellulose, and 20.58%-33.52% hemicellulose [6]. High linin and cellulose contents make EFB is difficult to decomposed. To change into simpler elements, EFB must be decomposed [7].

Various efforts to use EFB have been done, but the potential effort to develop by people surrounding oil palm factory is using EFB for straw mushroom planting media and compost fertilizer. EFB as organic or compost fertilizer was studied elsewhere [8,9], while EFB as planting media for straw mushroom was studied by [2].

Lampung province have studied the use EFB for straw mushroom media commercially. This is because the use of EFB is more beneficial than using rice straw. Amongst EFB benefits compared to rice straw include available in huge amount, cheaper price, and sustainable supply. Therefore, further research on the straw mushroom business potential by using EFB material in Lampung province is required, especially in the aspect of raw material.

The objectives of this research were to analyse the EFB production in Lampung Province; to analyse biology efficiency ratio of EFB to produce straw mushroom in a production scale, and to analyse the potential of straw mushroom by using EFB material in Lampung province.

## 2. Materials and methods

Research data were primary and secondary data. Primary data were collected from laboratory and field researches. Laboratory research was used to determine EFB characteristics including contents of lignin, cellulose, hemicellulose, water, and fat. Primary data used to determine straw mushroom potentials were biology efficiency ratios collected from results of field research, interviews with business actors, suppliers, collecting traders, academicians, and straw mushroom business practitioners.

The analysis of biology efficiency ratio in production scale was done in three districts. They were mushroom business units in South Lampung district, East Lampung district, and Middle Lampung district. The specifications of straw mushroom housing in the industrial scale of this research can be seen in Figure 1. Process stages of straw mushroom making by using EFB material can be seen in Figure 2.

Secondary data can be obtained from various forms of publications including scientific journals, statistical data especially presented on *Lampung Province in Numbers Y2015, Y2016 and Y2017* respectively [10-12]. Research data were tabulated and analysed and presented in figures and tables.

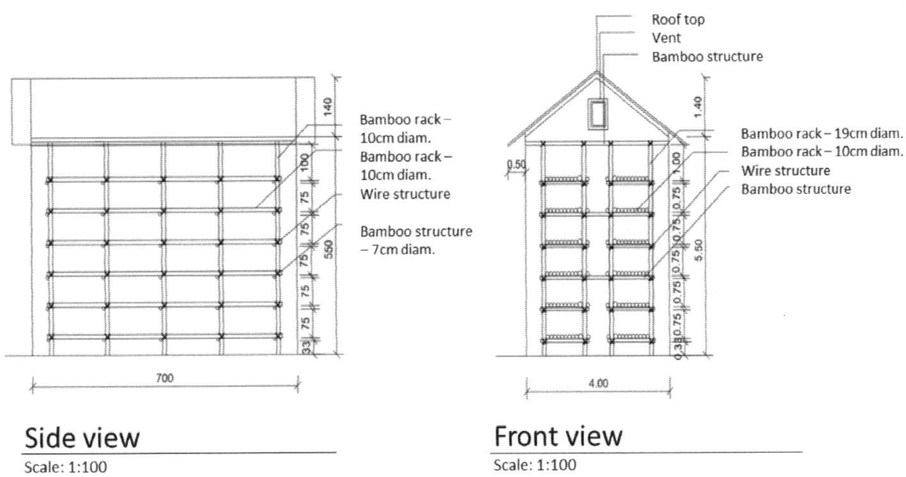


**Figure 1.** Specifications of Mushroom Housing for Research.

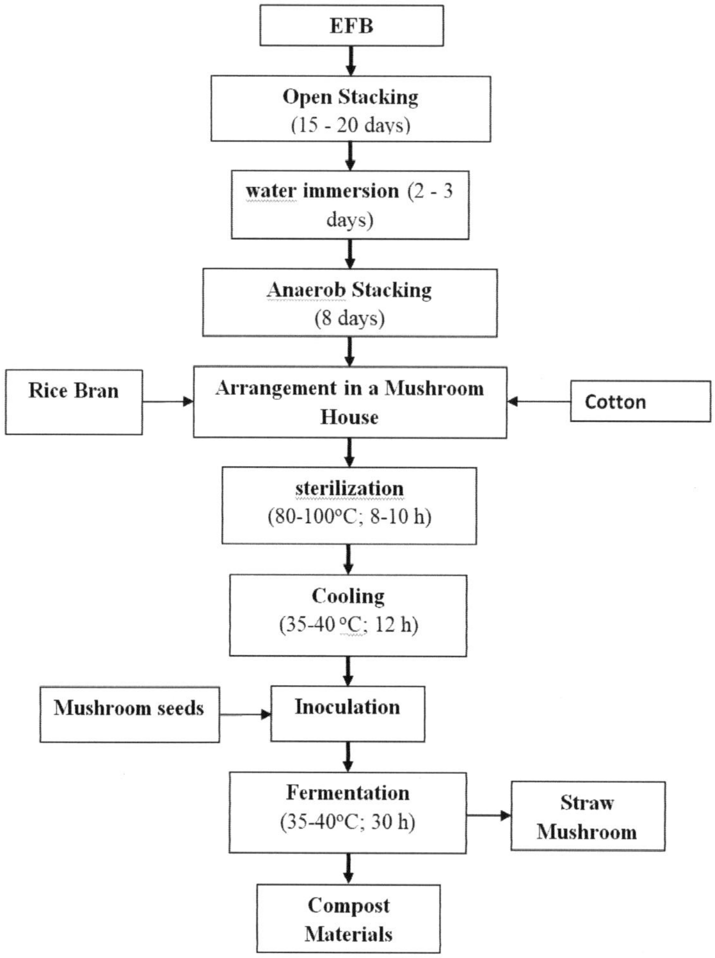


**Figure 2.** Stages of Production Process of Straw Mushroom from EFB.

## 3. Results and discussions

### *3.1. Productions of FFB and EFB in Lampung Province*

Table 1 shows the FFB production and potential of EFB in Lampung province based on the FFB production data from each district. The table shows that during successive three years (2015 to 2017), the FFB production fluctuated. The production fluctuation was caused by seasonal factor; at long dry season the FFB production declined [13]. Other factors were harvesting area expansion and plant maintenance. The biggest FFB producers in Lampung province in order were Mesuji district with average production of 173,248 ton, Middle Lampung district with average production of 106,536 ton, and Tulang Bawang district with average production of 70,171 ton. The percentages of FFB productions from each district in Lampung province can be seen in Figure 3.

Table 1 also shows potential of EFB production in Lampung province, where the EFB value depends on the amount of produced FFB. Wicke *et.al.* (2008) suggests that the amount of FFB product is between 16.0% to 23.0% from FFB, and 23.0% averagely [5]. An oil palm factory with a processing capacity of 60 ton/hour, 10 hours/day operation time, and 300 working days annually, will produce 41,400 tons of EFB [14].

About 23% of EFB from the total waste will cause environmental problem for industry and public if it is not managed seriously. The optimal management and use of EFB can improve competitiveness and productivity of oil palm industry [15].

**Table 1.** Production of FFB and Potential of EFB in Lampung province (2015-2017).

| No | Districts | FFB | | | EFB (23% from FFB)* | | |
|---|---|---|---|---|---|---|---|
| | | 2015 | 2016 | 2017 | 2015 | 2016 | 2017 |
| 1 | Lampung Barat | 6.333 | 4.900 | 5.617 | 1.457 | 1.127 | 1.292 |
| 2 | Tanggamus | 36 | 37 | 38 | 8 | 9 | 9 |
| 3 | Lampung Selatan | 35.331 | 40.197 | 42.630 | 8.126 | 9.245 | 9.805 |
| 4 | Lampung Timur | 8.897 | 8.659 | 8.540 | 2.046 | 1.992 | 1.964 |
| 5 | Lampung Tengah | 90.589 | 109.725 | 119.293 | 20.835 | 25.237 | 27.437 |
| 6 | Lampung Utara | 19.757 | 8.772 | 14.265 | 4.544 | 2.018 | 3.281 |
| 7 | Way Kanan | 41.617 | 27.507 | 34.562 | 9.572 | 6.327 | 7.949 |
| 8 | Tulang Bawang | 95.548 | 44.793 | 70.171 | 21.976 | 10.302 | 16.139 |
| 9 | Pasawaran | 3.172 | 2.325 | 2.749 | 730 | 535 | 632 |
| 10 | Pringsewu | 1.562 | 1.566 | 1.568 | 359 | 360 | 361 |
| 11 | Mesuji | 218.238 | 128.258 | 173.248 | 50.195 | 29.499 | 39.847 |
| 12 | Tulang Bawang Barat | 10.732 | 6.238 | 3.991 | 2.468 | 1.435 | 918 |
| 13 | Pesisir Barat | 14.379 | 20.282 | 23.234 | 3.307 | 4.665 | 5.344 |
| 14 | Bandar Lampung | 48 | 129 | 170 | 11 | 30 | 39 |
| 15 | Metro | 3 | 2 | 2 | 1 | 0 | 0 |
| | | 546.242 | 403.390 | 500.074 | 125.636 | 92.780 | 115.017 |

* From other sources [5]

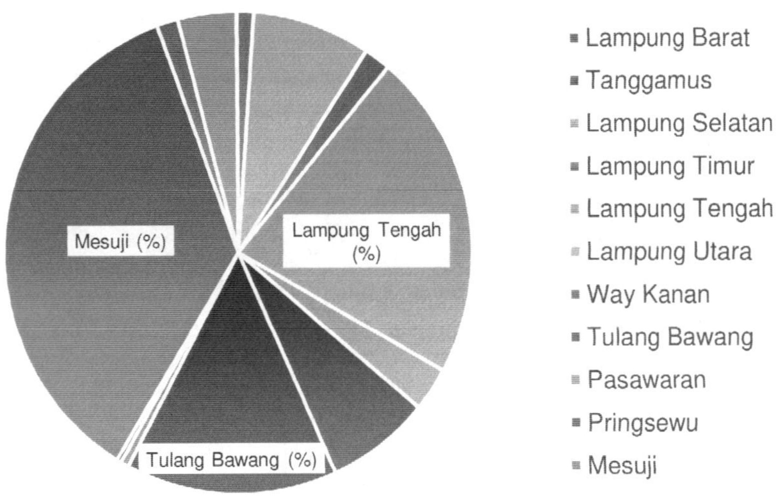


**Figure 3.** Percentage of FFB Production from each district in Lampung Province.

*3.2. Characteristics of EFB in Lampung Province*

The analysis result of EFB compositions in Lampung province compared to some other research results can be seen in Table 2.

**Table 2.** The Results of EFB Analysis and Analysis Result Data of EFB from Various Sources.

| No | Characteristics | Lampung Selatan | Lampung Timur | Lampung Tengah | Chan (2014) | [5] |
|---|---|---|---|---|---|---|
| 1 | Cellulose (%) | 36,19 | 36,45 | 37,50 | 23,70 – 65,00 | 43,30 |
| 2 | Hemicellulose (%) | 18,84 | 16,08 | 22,44 | 20,58 – 33,52 | 26,20 |
| 3 | Lignin (%) | 22,75 | 24,73 | 23,03 | 14,10 – 30,45 | 30,50 |
| 4 | Water (%) | 7,41 | 5,25 | 5,08 | 2,40 – 14,28 | N. A |
| 5 | Fat (%) | 5,29 | 4,62 | 5,63 | N. A | N. A |

The table shows that three biggest and important components in EFB are cellulose, hemicellulose, and lignin. Table 2 also shows that cellulose, hemicellulose, and lignin components of EFB from one district compared to other districts are almost the same. The compositions of cellulose, hemicellulose, and lignin of EFB are influenced by plant age, FFB maturity level, plant varieties, and locations of oil palm planting [16]. The use or processing of EFB must pay attention to these three component characteristics.

Cellulose is a substance composing plant that is found in cellular structure. The cellulose content in the plant cellular walls is always higher compared to lignin and hemicellulose in dry weight of a plant. Cellulose is a glucose polymer with ß-1binding, 4 glucosides in a straight chain [16]. The crystalline cellulose structure causes cellulose difficult to decompose properly both acid and basal ways. In order to decompose enzymatically, lignocellulose must be delignified to facilitate cellulose enzyme work in decomposing cellulose [17].

Lignin is a plant composing component together with cellulose and other fiber materials to form plant structure and cellular parts. In the plant stem, lignin function as binding material of other composing components, so that a plant is able to stand straight. Lignin is formed from aromatic clusters that are interconnected with affiliated chains that contain of 2-3 carbons. In a pyrolysis process of lignin, aromatic chemical compound will be produced in form of phenol, especially cresol [16].

Lignin structure is very hard to decompose chemically and enzymatically. Lignin, cellulose, and hemicellulose can always be found in plant cellular wall structures, so that they are often categorized as carbohydrate, even though lignin is not carbohydrate. This is proven by the higher proportion of lignin [18]. The lignin contents differ between one plant to another, for example, lignin contents were 22.65 in pine straw, 20.40% in wheat straw, and 14.88% in ramie fiber [19].

Hemicellulose is a group of heterogenous polysaccharide with lower molecular weight compared to cellulose. Hemicellulose amount is about 15 to 30 percent of lignocellulose dry weight material. Hemicellulose is relatively easier to hydrolyse with acid into monomers containing glucose, mannose, galactose, xylose, and arabinose. Hemicellulose also is crossed-binding with lignin to form complex networks and to provide strong structures. Hemicellulose contains of units of D-glucose, D-galactose, D-mannose, D-xylose, and L-arabinose that are formed simultaneously in varying glycosylic binding and combinations [19].

### *3.3. Biology efficiency ratio of EFB material-based straw mushroom*

Another important parameter is biology efficiency ratio functioning to measure substrate efficiency use level. The higher is the biology efficiency ratio, the more efficient will be the substrate use by the straw mushroom [20]. Compare to others stated that biology efficiency ratio is the produced fresh straw mushroom weight to divide by substrate material (EFB) and multiplied by 100% [21].

The research results from three EFB material-based straw mushroom business units showed that the biology efficiency ratio of EFB material-based straw mushroom differed from one business unit to another (Figure 4). The biology efficiency value of EFB conversion into straw mushroom was about 3.00% to 4.81%. The research result by other showed that EFB could be used as white oyster mushroom planting media to replace wood sawdust with biology efficiency ratio value of more than 30% [20-21].

Also, Purindraswari and Udiantoro (2016) [21] showed that the use of media by composition of 750 gr EFB and 250 gr rice straw could produce an average of 19 straw mushrooms with 60.88 gr of weight or a biology efficiency ratio of 6.088%. The composition of 250 gr of EFB and 750 gr rice straw could produce an average of 2 straw mushroom with 5.72 gr of weight or a biology efficiency ratio of 0.572%. Many factors influenced conversion values including mushroom seed quality, climate surrounding mushroom housing, mushroom housing maintenance, and nutrition composition in the planting media media [21].

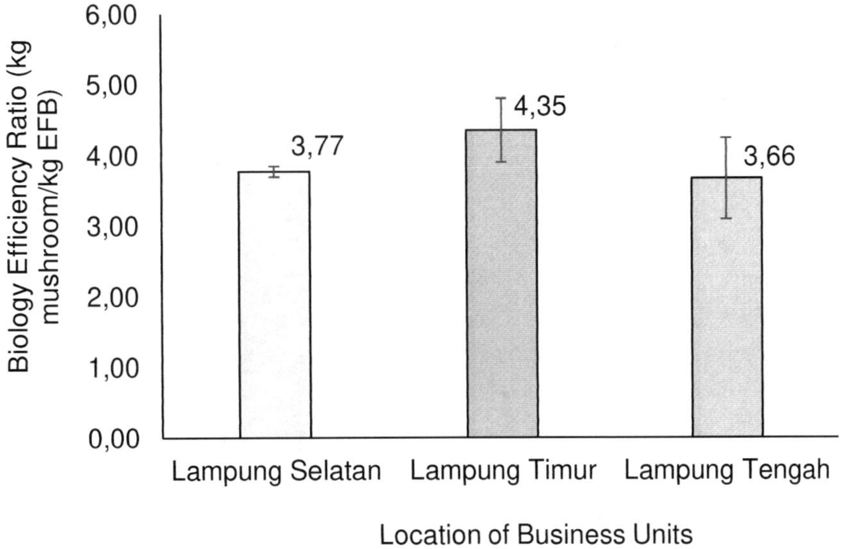


**Figure 4.** Biology efficiency ratio of EFB material-based straw mushroom.

Figure 4 shows that the highest EFB material-based straw mushroom efficiency is in East Lampung. This was because management in East Lampung was better compared to other districts. EFB material-based straw mushroom cultivation in East Lampung was conducted professionally, started from the experienced human resource, bigger business scale, well-established marketing and business networks [22].

Many factors influence the biology efficiency ratio value of straw mushroom. They are mushroom seed quality, planting media thickness, temperature, RH, and used media nutrition [21]. In this research, planting media thickness was 30-40 cm, temperature was 35-50$^0$C, RH was about 90-95%, and rice bran was added for additional nutrition, while some business units added cotton to maintain RH.

### *3.4. EFB material-based straw mushroom potential in Lampung Province*

Table 1 shows FFB production and EFB potential in Lampung province for three successive years from 2015 to 2017. EFB production was very dependent to amount of FFB production. Some sources said that 20-25% EFB was produced from FFB and the average was 23% [5]. EFB potential in Lampung province in 2017 reached 115,017 ton with a production average of 111,144 ton annually.

The field observation results showed that from 15 districts/municipals in Lampung province, only 5 districts had oil palm factories; 6 factory units in Middle Lampung district, 1 factory unit in North Lampung district, 2 factory units in Way Kanan district, 3 factory units in Tulang Bawang district, and 5 factory units in Mesuji district. Therefore, FFB coming from districts that do not have oil palm factory is sent to those districts with oil palm factory.

This research showed that considerations concerning FFB deliveries from a district to another were determined by distances between plantations to the oil palm factories, access into the oil palm factories, and FFB prices. FFB distributions from districts into other districts with oil palm factories can be seen in Figure 5. Amounts of FFB sent from district to other districts with oil palm factories can be seen in Table 3.

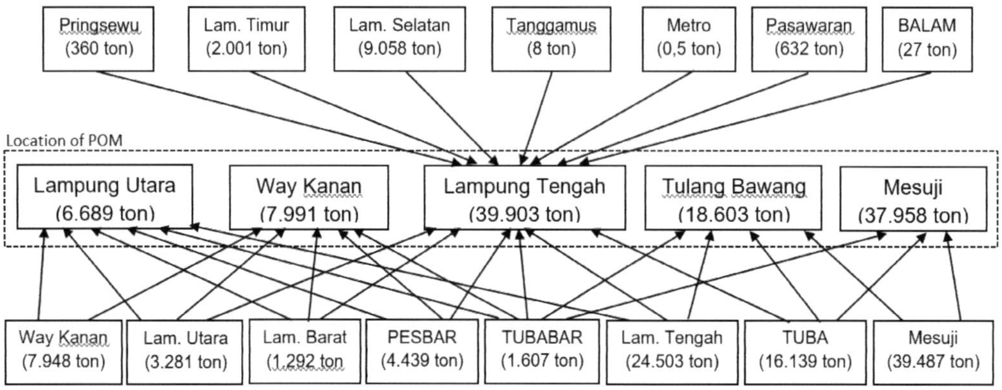


**Figure 5.** Distribution and transportation of FFB from districts to other districts with oil palm factories in Lampung Province.

**Table 3.** Amount of FFB sent from districts to other districts with oil palm factories.

| Regency / City | TBS (Ton) | District that has a Palm Oil Mill (POM) | | | | | | | | | |
|---|---|---|---|---|---|---|---|---|---|---|---|
| | | Lampung Tengah | | Way Kanan | | Lampung Utara | | Tulang Bawang | | Mesuji | |
| | | (%) | (Tons) | (%) | (Tons) | (%) | (Tons) | (%) | (Tons) | (%) | (Tons) |
| Lampung Barat | 5.617 | 50% | 2.808 | 20% | 1.123 | 30% | 1.685 | - | - | - | - |
| Tanggamus | 37 | 100% | 37 | - | - | - | - | - | - | - | - |
| Lampung Selatan | 39.386 | 100% | 39.386 | - | - | - | - | - | - | - | - |
| Lampung Timur | 8.699 | 100% | 8.699 | - | - | - | - | - | - | - | - |
| Lampung Tengah | 106.536 | 90% | 95.882 | - | - | 5% | 5.327 | 5% | 5.327 | - | - |
| Lampung Utara | 14.265 | 10% | 1.426 | 5% | 713 | 85% | 12.125 | - | - | - | - |
| Way Kanan | 34.562 | 10% | 3.456 | 80% | 27.650 | 10% | 3.456 | - | - | - | - |
| Tulang Bawang | 70.171 | 10% | 7.017 | - | - | - | - | 80% | 56.136 | 10% | 7.017 |
| Pasawaran | 2.749 | 100% | 2.749 | - | - | - | - | - | - | - | - |
| Pringsewu | 1.565 | 100% | 1.565 | - | - | - | - | - | - | - | - |
| Mesuji | 173.248 | - | - | - | - | - | - | 10% | 17.325 | 90% | 155.923 |
| Tulang Bawang Barat | 6.987 | 10% | 699 | 20% | 1.397 | 10% | 699 | 30% | 2.096 | 30% | 2.096 |
| Pesisir Barat | 19.298 | 50% | 9.649 | 20% | 3.860 | 30% | 5.789 | - | - | - | - |
| Bandar Lampung | 116 | 100% | 116 | - | - | - | - | - | - | - | - |
| Metro | 2 | 100% | 2 | - | - | - | - | - | - | - | - |
| Number of FFB / Reg. (tons) | 483.235 | | 173.491 | | 34.743 | | 29.081 | | 80.884 | | 165.036 |
| Production of EFB (tons) | | | 39.903 | | 7.991 | | 6.689 | | 18.603 | | 37.958 |

The determination of EFB material-based straw mushroom potential was based on the research on biology efficiency ratio from EFB into straw mushroom, which was 4.35%, and the amount of EFB produced from FFB. The analysis of EFB material-based straw mushroom in Lampung province can be seen in Figure 6.

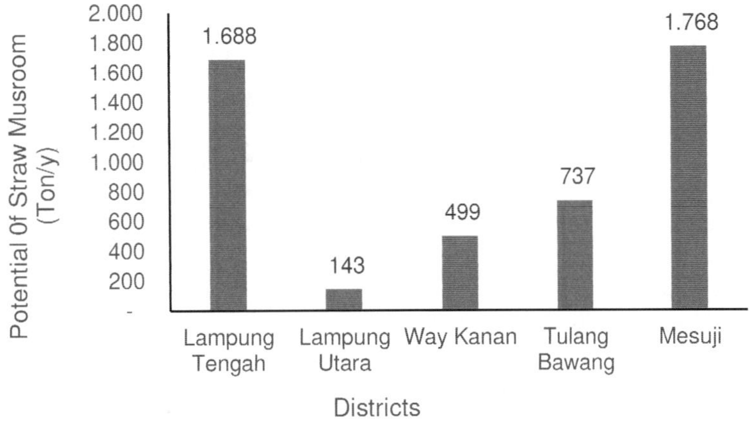


**Figure 6.** The EFB material-based straw mushroom potential in Lampung Province.

## 4. Conclusion

EFB production Lampung province reached 111,144 ton annually and it was not yet used optimally. The biology efficiency ratio of EFB into straw mushroom in production scale was 3.93%. The potential of EFB material-based straw mushroom business development in Lampung province was 4,835 ton annually; they were 1,768 ton from Mesuji district, 1,688 ton from Middle Lampung district, 737 ton from Tulang Bawang district, 499 ton from Way Kanan district, and 143 ton from North Lampung district.

## References

[1] Das S, S-H Lee, P Kumar, K-H Kim, S S Lee, and S S Bhattacharya 2019 Solid waste management: Scope and the challenge of sustainability, *J. Clean. Prod.* 658–678. https://doi.org/https://doi.org/10.1016/j.jclepro.2019.04.323.

[2] Wang X, P Zhou, J Cheng, Z Chen, and X Liu 2018 Use of straw mushrooms (Volvariella volvacea) for the enhancement of physicochemical, nutritional and sensory profiles of Cantonese sausages., Meat Sci. 18–25. https://doi.org/https://doi.org/10.1016/j.meatsci.2018.07.033.

[3] Julian A V, R G Reyes, and F Eguchi 2018 Agro-Industrial Waste Conversion Into Medicinal Mushroom Cultivation, Encycl. Environ. Heal. 13–20Agro-Industrial Waste Conversion Into Medicinal Mushroom Cultivation, in: Encycl. Environ. Heal., (Second Ed, Elsevier B.V., : pp. 13–20.

[4] Liew W L, M A Kassim, K Muda, S K Loh, and A C Affam 2015 Conventional methods and emerging wastewater polishing technologies for palm oil mill effluent treatment: A review, J. Environ. Manage. https://doi.org/https://doi.org/10.1016/j.jenvman.2014.10.016.

[5] S S, P Kongjan, and S O-Thong 2017 Biogas Production from Anaerobic Co-digestion of Palm Oil Mill Effluent and Empty Fruit Bunches., Energy Procedia. 717–722. https://doi.org/https://doi.org/10.1016/j.egypro.2017.10.206.

[6] Zhang Z H, J Chen, L Li, M Tao, C Zhang, Q B Qin, J Xiao, Y Liu, and S J Liu 2014 Research advances in animal distant hybridization, Sci. China. Life Sci. 57: 889–902. https://doi.org/10.1007/s11427-014-4707-1.

[7] Harun N H H M, N A F A Samad, and S Saleh 2017 Development of Kinetics Model for Torrefaction of Empty Fruit Bunch from Palm Oil Waste, Energy Procedia. 744–749. https://doi.org/https://doi.org/10.1016/j.egypro.2017.03.385.

[8] J.J. T, M D Zulkali, and M A Hashim 1995 Microbiological and biochemical changes during the composting of oil palm empty-fruit-bunches. Effect of nitrogen supplementation on the substrate., Bioresour. Technol. 52: 133–144. https://doi.org/https://doi.org/10.1016/0960-8524(95)00008-3.
[9] Yuvaneswaran Krishnan, C P C Bong, N F Azman, Z Zakaria, N A Othman, N Abdullah, C S Ho, C T L S B Hansen, and H Hara 2017 Co-composting of palm empty fruit bunch and palm oil mill effluent: Microbial diversity and potential mitigation of greenhouse gas emission., J. Clean. Prod. 94–100. https://doi.org/https://doi.org/10.1016/j.jclepro.2016.08.118.
[10] BPS-Statistics of Lampung Province 2016 Lampung Province In Figures 2015,Lampung Province In Figures 2015, BPS-Statistics of Lampung Province, .
[11] BPS-Statistics of Lampung Province 2017 Lampung Province In Figures 2016,Lampung Province In Figures 2016, BPS-Statistics of Lampung Province, .
[12] BPS-Statistics of Lampung Province 2018 Lampung Province In Figures 2017,Lampung Province In Figures 2017, BPS-Statistics of Lampung Province, .
[13] Dian Afriyanti, C Kroeze, and A Saad 2016 Indonesia palm oil production without deforestation and peat conversion by 2050., Sci. Total Environ. 557–558: 562-570. https://doi.org/https://doi.org/10.1016/j.scitotenv.2016.03.032.
[14] Nasution M A, T Herawan, and M Rivani 2014 Analysis of Palm Biomass as Electricity from Palm Oil Mills in North Sumatera, Energy Procedia. 166–172. https://doi.org/https://doi.org/10.1016/j.egypro.2014.01.210.
[15] Nasution M A, D S Wibawa, T Ahamed, and R Noguch 2018 Comparative environmental impact evaluation of palm oil mill effluent treatment using a life cycle assessment approach: A case study based on composting and a combination for biogas technologies in North Sumatera of Indonesia, J. Clean. Prod. 1028–1040. https://doi.org/https://doi.org/10.1016/j.jclepro.2018.02.299.
[16] D W, M Nuruddin, M Hosur, A Tcherbi-Narteh, and S Jeelani 2015 Extraction and characterization of lignin from different biomass resources., J. Mater. Res. Technol. 4: 26–32. https://doi.org/https://doi.org/10.1016/j.jmrt.2014.10.009.
[17] Mosier N, C Wyman, B Dale, R Elander, Y Y Lee, M Holtzapple, and M Ladisch 2005 Features of promising technologies for pretreatment of lignocellulosic biomass, Bioresour. Technol. 96: 673–686. https://doi.org/https://doi.org/10.1016/j.biortech.2004.06.025.
[18] M T-G, R Panahi, and M Babak 2019 Optimizing the combination of conventional carbonaceous additives of culture media to produce lignocellulose-degrading enzymes by Trichoderma reesei in solid state fermentation of agricultural residues., Renew. Energy. 946–955. https://doi.org/https://doi.org/10.1016/j.renene.2018.07.130.
[19] Karimi K, and M J Taherzadeh. 2016 A critical review on analysis in pretreatment of lignocelluloses: Degree of polymerization, adsorption/desorption, and accessibility, Bioresour. Technol. 348–356. https://doi.org/https://doi.org/10.1016/j.biortech.2015.12.035.
[20] Liang C H, C-Y Wu, P-L Lu, Y-C Kuo, and Z-C Liang 2019 Biological efficiency and nutritional value of the culinary-medicinal mushroom Auricularia cultivated on a sawdust basal substrate supplement with different proportions of grass plants, Saudi J. Biol. Sci. 26: 263–269. https://doi.org/https://doi.org/10.1016/j.sjbs.2016.10.017.
[21] Loss E, A R Royer, M Barreto-Rodrigues, and A C Barana 2009 Use of maize wastewater for the cultivation of the Pleurotus spp. mushroom and optimization of its biological efficiency, J. Hazard. Mater. 166: 1522–1525. https://doi.org/https://doi.org/10.1016/j.jhazmat.2008.11.014.
[22] Sarono S, and E I Siregar 2018 Analysis of Business Development Strategy Straw Mushroom Raw Material EFB in Lampung Province, Int. Conf. Appl. Sci. Technol.Analysis of Business Development Strategy Straw Mushroom Raw Material EFB in Lampung Province, in: Int. Conf. Appl. Sci. Technol., Manado, .

# Sensitivity test of electromagnetic sensor using closed type transverse electromagnetic (TEM) cell

**H H Sinaga[1,*], R B Vraja[1], H B H Sitorus[1], D Permata[1] and Y Yuniati[1]**

[1] Department of Electrical Engineering, Universitas Lampung, Jalan Prof. Soemantri Brojonegoro No. 1, Bandar Lampung 35145, Indonesia

*Email: herman.h.sinaga@eng.unila.ac.id

**Abstract.** Electromagnetic method is one of the proved successfully applied to detect the presence of partial discharge events on electrical insulation. This method applied by utilized an electromagnetic sensor that capable to capture the electromagnetic signals generated by the partial discharge source. The sensors need high sensitivity to be able to capture the very small electromagnetic signals. Thus, the sensors need to be tested its sensitivity to make sure its ability to capture the electromagnetic sensor generated by the partial discharge source. In this research, a Transverse Electromagnetic (TEM) cell is applied to test the sensor sensitivity. The TEM cell is a close cell type which provide better noise insulation and has frequency response up to 1 GHz. The experiment results show the TEM cell can be used to test the sensor sensitivity. The sensor sensitivity is much depending on the sensor dimension. The larger the sensor the higher its sensitivity. Future research should be arranged to test the sensor sensitivity to capture wide bandwidth PD signals.

**Keywords:** Sensitivity Test, Partial discharge, Electromagnetic sensor, TEM cell

## 1. Introduction

Partial discharge (PD) events in electrical insulation will produce electromagnetic signals that transverse in all direction outward the partial discharge sources. So, to be able to detect the presence of the partial discharge, one can utilize the electromagnetic signals by using appropriate method. Method to detect the presence of the PD in electrical insulation is known as electromagnetic method. This method applied to detect the electromagnetic signals generated by the PD sources by using an antenna which is acts as electromagnetic sensor [1,2]. Other researchers have pointed out the electromagnetic sensors to detect the presence of the PD's [1-5]. The electromagnetic method has proven its ability to detect PD in varying type electrical insulation such as transformer [2] and gas insulated switchyard (GIS) [6,7]. However, there is a drawback of the using of the electromagnetic method, i.e. it is difficult if not impossible to calibrate. To convert the electromagnetic measurement results is almost impossible to do due to the nature of the electromagnetic signals itself [8,9]. The electromagnetic signals will travel all-around of the PD source thus the signals will attenuate along its path. This will cause the magnitude of the electromagnetic signals heavily depends on the electromagnetic signal path [9] and not the distance of the sensor to the PD sources [10]. Thus, without knowing the exact location of the PD source, the magnitude of the electromagnetic signals cannot be converted to a more convenient unit charge coulomb [9].

To override the calibration inability, CIGREE TF 15/33.03.05 [11] propose sensitivity test for the electromagnetic sensor. The sensor sensitivity test conduct inside a uniformly electromagnetic chamber, which commonly known as transverse electromagnetic (TEM) cell. In this paper, discussed the using of a closed type TEM cell to test the sensor sensitivity.

## 2. Sensor Sensitivity Test

In every measurement system, calibration which compares the output of a piece of measuring equipment to a standard value is required. The calibration establishes with certainty the amount being measured. However, this is not the case for the electromagnetic detection method. The output of the electromagnetic sensors cannot be calibrated as per IEC 60270 as they do not directly quantify the amount of charge of the PD pulses [11].

The reason is that a PD can occur at almost any location inside the transformer tank. The path of the electromagnetic signals from the PD source to the sensor is affected by the structure inside the transformer. The PD signal propagation can be obstructed by some solid material parts inside the transformer. The active parts of the transformer also affect the attenuation of the electromagnetic signals which caused the attenuation not linear to the distance. Thus, without knowing the exact location of the PD, it is difficult to convert the amount of PD detected by the electromagnetic sensor to an equivalent pC level [10].

To provide information about the sensor capabilities, it is important to set up tests which are repeatable and can be used to test various sensors. In [10,12] sensor calibration is introduced and information is provided about the sensor frequency response. A μ-TEM cell was used in [13] to test antenna response. In [14, 15], a TEM cell was built to test the sensor frequency response in an attempt to calibrate the sensor for PD detection. The TEM cell can also be used to test the sensor response for specific pulses such as the step pulse [16] in order to find the most suitable sensor for application of PD diagnostic and monitoring. Using a step pulse to determine the frequency response of the sensors has an advantage over the sweep frequency generator [16]. This is because the step pulse contains all necessary frequency components so only one measurement is needed. Also, the cost of the test can be reduced as the frequency generator can be eliminated. However, the sweep frequency generator will provide real frequency response where input and output can be compared directly, unlike the step pulse where the pulse signals must be converted to frequency response.

As the electromagnetic sensors cannot be calibrated based on a pC value, CIGRE TF 15/33.03.05 has recommended a method for its sensitivity verification which can be used to determine on-site the minimum sensitivity of this measuring method in GIS [11]. The sensitivity test will show the amount of PD which can be measured by the electromagnetic method. The sensitivity of the electromagnetic method is very dependent on the type of sensor, types of PD source and the surrounding structure [17].

### *2.1. UHF Electromagnetic Signal*

The electromagnetic pulses generated by the PD source can carry a wide band frequency signal depending on their rise time. In air, the propagation velocity of the electromagnetic signals is approximately as fast as the speed of light (*c*). In other media, the speed of the electromagnetic signal (*v*) depends on the permittivity and permeability of the material. This factor is called the refractive index of the material and is expressed as:

$$\eta = \frac{c}{v} = \sqrt{\frac{\mu\varepsilon}{\mu_o\varepsilon_o}} = \sqrt{\mu_r\varepsilon_r} \qquad 1$$

where $\varepsilon = \varepsilon_r\varepsilon_o$ is the permittivity of the material, and $\mu = \mu_r\mu_o$ is the permeability of the material (expressed as a product of its relative value and the absolute value of free space). In a vacuum, the propagation velocity of electromagnetic signals is the speed of light, i.e. 3 x $10^8$ m/s.

IOP Conf. Series: Materials Science and Engineering **857** (2020) 012018 doi:10.1088/1757-899X/857/1/012018

### *2.2. Sensor sensitivity*

The placement of the UHF sensors to detect PD events in transformers is usually at fixed locations. The locations of sensor installation depend on the type of sensor and transformer construction, as described in a previous section. As the PD can occur in any location in the transformer, the electromagnetic signal path from the PD source to the sensor can be affected by the structure inside the transformer. The signal path is also affected by the density of the insulation oil in the transformer tank which can also be subjected to variations in temperature (caused by loading changes). In addition, the live active parts of the transformer can obstruct electromagnetic signals and cause signal attenuation whereby the signal becomes not linear to the distance. This will cause conversion of the signal magnitude detected by the UHF sensor to an equivalent pC level to become difficult [10].

CIGRÉ WC TF 15/33.03.05 [11] recommended sensitivity verification for UHF method as a substitution for calibration. Sensitivity verification can be used on-site to determine the minimum sensitivity of the measuring system. Although this recommendation is only for GIS, with some adjustment this method can also be applied to power transformers which use oil insulation. The sensitivity of the UHF measuring method is very dependent on the type of sensor, on the type of the PD defect, and on the location of the PD source [17].

## 3. Transverse electromagnetic cell (TEM cell)

TEM cell is used to provide a testing chamber for electrical apparatus, whenever the testing procedure requires uniform electromagnetic around the testing sample. Also, the TEM cell, provide barrier from unwanted electromagnetic signals that probably obstruct measuring results.

The basic TEM cell consists of a parallel conductor plate as shown in Figure 1. One conductor is wider and connected to ground. The top conductor with specific widths is arranged with specific distance to produce desire impedance. In this paper, discussed the TEM cell with impedance 50 Ohms. The parallel conductor TEM cell as shown in figure 1 is the simplest form of the TEM cell. This type has disadvantage due to the presence of unwanted (noise) signals.

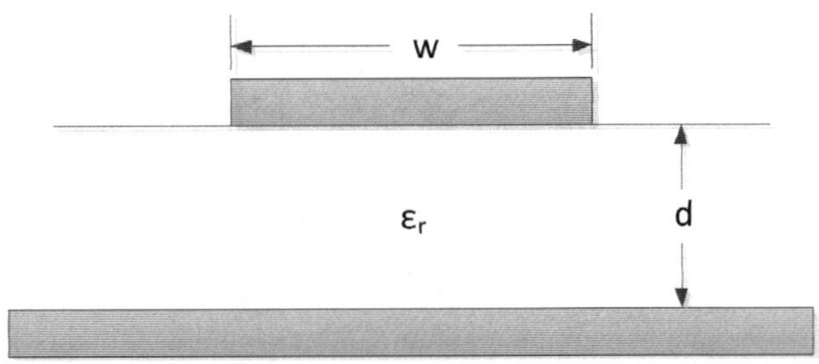


**Figure 1**: Cross section of a parallel conductor of TEM cell.

IOP Conf. Series: Materials Science and Engineering **857** (2020) 012018 doi:10.1088/1757-899X/857/1/012018

**Figure 2.** Closed type TEM cell use to test sensor sensitivity.

To overcome the open type TEM cell disadvantage, a close type TEM cell is used in this paper. The close type TEM cell used in this paper is shown in figure 2. The TEM cell built using a dual layer PCB.

## 4. Result and discussion

The experiment setup diagram is shown in Figure 3 and arrangement is shown in Figure 4. The step pulse input signals produce by a function generator. Step pulse input connected to one port of the TEM cell. A dual channels oscilloscope is used to record the step pulse on the input port and the sensor output. The other end port of the TEM is connected to a 50-ohm termination. Two type of sensors use in experiment i.e. monopole and a planar-spectral sensor. The Monopole sensor has 5 cm length and the spectral sensors have two size, that is 5 cm and 10 cm wide.

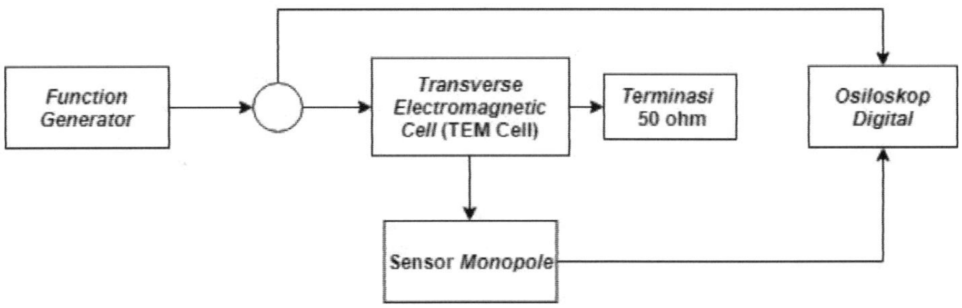


**Figure 3**. Block diagram sensor sensitivity test.

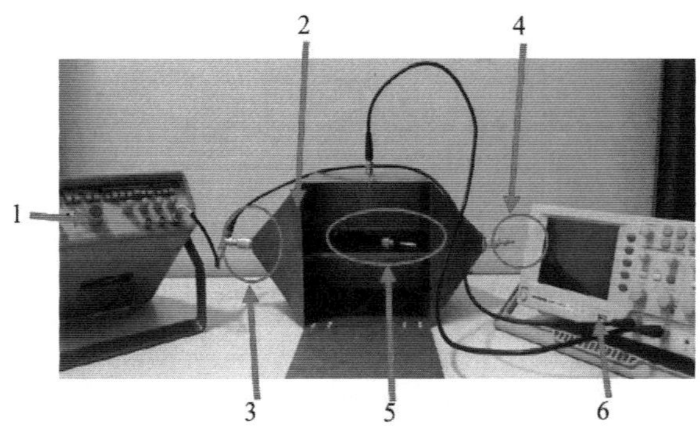


**Figure 4.** Sensitivity test of the electromagnetic sensor using TEM cell.
*1. Function generator, 2. Transverse electromagnetic* (TEM) *cell*, *3.* T BNC *connector, 4.* A 50-ohm termination, 5. *Monopole or spectral sensor and 6.* Oscilloscope

The sensor signal responses for each type of sensors are shown in Figure 5. The step pulse input shown in red color and the sensor response in blue color. The three sensors have quite similar response, with fast oscillation after the step input signal hit the TEM cell input port. The time delay between the step input signals and the sensors output seems undetectable. This might due to the length of the TEM is quite small. Thus, the time needed by the step pulse signals from the input port to reach the other end port is almost instantaneous.

IOP Conf. Series: Materials Science and Engineering **857** (2020) 012018 doi:10.1088/1757-899X/857/1/012018

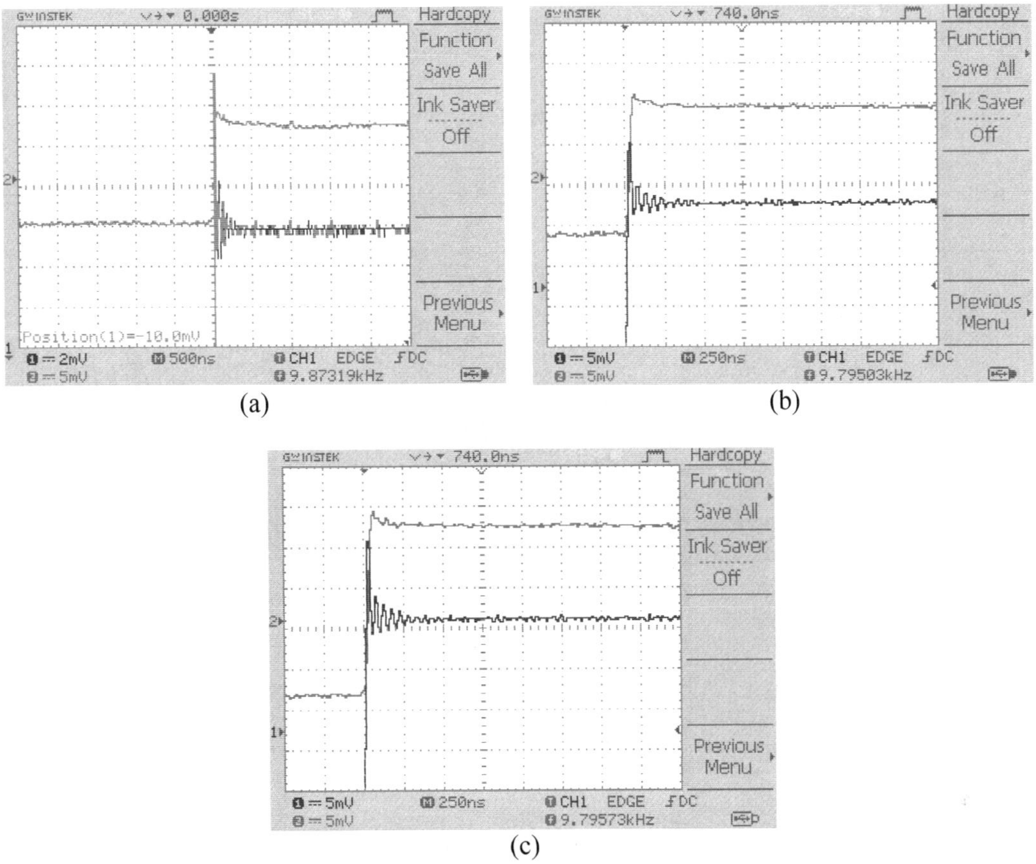


**Figure 5.** Typical sensor response (a) monopole (b) fractal 5 cm (c) fractal 10 cm.

Of the three sensors, all sensor has ability to response to the step pulse signals. The fractal sensor with size 10 cm has higher signals response as it has bigger dimension compare to monopole and 5 cm spectral sensor. The monopole has lower waveform magnitude and longer oscillation. Nevertheless, the three sensors have sensitivity to detect the step pulse thus should be able to detect electromagnetic signals generated by a PD source.

## 5. Conclusion

The experiment result shows the TEM cell can be used to test the sensitivity of the electromagnetic sensor. Two types of sensor are tested in this paper, i.e. monopole and spectral sensor. The two types sensor show different sensitivity, with spectral sensor has highest sensitivity. The larger the spectral sensor size, the higher its sensitivity. The sensitivity test using TEM cell provide data about the sensor ability to detect the electromagnetic signal, thus should be able to detect the electromagnetic signals generated by a partial discharge source. Further research is to compare the sensitivity test results to the ability of the sensor to detect electromagnetic signals produce by partial discharge events. The sensor ability to detect the signals in wide frequency bandwidth should be studied. The frequency range of the sensor works can be measure using a spectrum analyzer. Input to the TEM cell also need to modify so the input signals contain a wide band.

**Acknowledgements**
The authors would like to express their gratitude to Kemenristek-DIKTI Indonesia for supporting this research. This research was supported by Kemenristek-DIKTI Indonesia with the Fundamental Research Grant, contract number: 857/UN26.21/PN/2019.

**References**
[1] Judd M D, Yang L and Hunter I B B 2005 *IEEE Electrical Insulation Magazine* **21** 2 pp 5-14.
[2] Judd M D, Yang L and Hunter I B B 2005 *IEEE Electrical Insulation Magazine* **21** 3 pp 5-13.
[3] Judd M D, Cleary G P, Bennoch C J, Pearson J S and Breckenridge T 2002 Electrical Insulation *Conference Record of the IEEE Int. Symp. on Electrical Insulation* pp 145-149.
[4] Sinaga H H, Phung B T and Blackburn T R 2009 *19th Australasian Universities Power Engineering Conference (AUPEC'09)* paper PP027.
[5] Sinaga H H, Phung B T and Blackburn T R 2012 *Int. Conf. on Condition Monitoring and Diagnosis (CMD2012)*
[6] Wenger P, Beltle M, Tenbohlen S, Riechert U, Behrmann G 2019 *IEEE Transactions on Power Delivery*, **34** 4 pp 1540 - 1548
[7] Bin F, Wang F, Sun Q, Lin S, Xie Y, Fan M 2018 *IET Microwaves, Antennas & Propagation* **12** 14 pp 2184 - 2190
[8] Stone G C 2005 *IEEE Transactions on Dielectrics and Electrical Insulation* **12** 5 pp 891-903
[9] Cavallini A, Montanari G C, Tozzi M 2010 *IEEE Transactions on Dielectrics and Electrical Insulation* **17** 1 pp: 198 – 205, 2010
[10] Coenen S, Tenbohlen S, Markalous S M, Strehl T 2008 *IEEE Transactions on Dielectrics and Electrical Insulation* **15** 6 pp 1553 – 1558
[11] CIGRE TF 15/33.03.05 1999 *Electra* **183** pp 75-87
[12] Judd M D and Farish O 1998 *IEEE Transactions on Instrumentation and Measurement* **47** 4 pp 875-880
[13] Kang N W, Kang J S, Kim D C and Kim J H 2007 *IEEE Transactions on Instrumentation and Measurement* **56** 2 pp 435-438
[14] Judd M D 1999 *IEEE Proceedings-Science, Measurement and Technology* 1**46** 3 pp 113-116
[15] Ishigami S and Hirata M 2009 *Proc. 20th Int. Zurich Symposium on EMC* pp 425-428
[16] Judd M D, Farish O and Pearson J S 1997 *IEE Proceedings - Science, Measurement and Technology* **144** 3 pp 117 - 122
[17] Sinaga H H, Phung B T, Ao A P and Blackburn T R 2011 *XVII Int. Symp. on High Voltage Engineering* Paper **D**-069.

IOP Conf. Series: Materials Science and Engineering **857** (2020) 012019 doi:10.1088/1757-899X/857/1/012019

# Methylene Blue Adsorption Isotherm on *Spirulina* sp. Microalgae Biomass Coated by Silica-Magnetite

**R A Kausar[1]*, Buhani[1], and Suharso[1]*,**

[1]Departement of Chemistry, Faculty of Mathematics and Natural Sciences, University of Lampung, Jalan Prof. Soemantri Brojonegoro No.1, Bandar Lampung, 35145, Indonesia

*E-mail: radhoalkausar@gmail.com; suharso@fmipa.unila,ac.id

**Abstract.** In this study, it has been performed modification of *Spirulina* sp. algae biomass with silica (HAS) coated with magnetite particles (HASM) and its adsorption test on a solution of methylene blue dye (MB) in solution. The identification of functional groups from HASM adsorbent was studied using infra-red (IR) spectrophotometer, the degree of crystallinity of the material was determined by X-ray diffraction (XRD), while the morphology and HASM constituent elements were analyzed by Scanning Electron Microscopy-Energy-Dispersive-X-Ray (SEM-EDX). The HASM adsorption isotherm of MB solution was studied by applying adsorption experiments using the batch method. The concentrations of MB dye in the adsorption process were analyzed by a UV-Vis spectrometer. Adsorption of 100 mg $L^{-1}$ MB dye solution by HASM adsorbent at pH of 6 with a contact time of 60 minutes resulted in an adsorbed MB amount of 90.90 mg $g^{-1}$. The adsorption isotherm model of the MB dye solution in the HASM adsorbent tends to follow the model of Freundlich adsorption isotherm. The adsorbent HASM is the effective adsorbent to absorb MB dye in solution.

**Keywords**: *Spirulina* sp., adsorption, silica coated magnetite.

## 1. Introduction

Increasing industrial development has an influence on the environment, such as diverting pollution from chemicals both containing organic and inorganic [1]. The main problem with this chemical pollution is the difficulty of decomposing these compounds, so that it will have a negative impact on the environment. One of the non-biodegradable organic compounds in the environment is textile dyes. Textile dyes are generally made from azo mixture and derivatives of benzene groups. The benzene groups are very difficult or need a long time to degrade. Azo compounds if too long in the environment, will be a source of disease because they are carcinogenic and mutagenic [2].

One textile dye that pollutes the environment is methylene blue. Methylene blue dyes are textile dyes made from azo substances and from azo benzene groups [3]. Therefore, it is necessary to do an effective alternative to reduce the concentration of MB from the remnants of dissolved dyes and also from impurities obtained from natural fibers to avoid environmental pollution more widely [4].

To reduce pollution, several techniques have been used, including coagulation, complexation, ion transfer, and adsorption techniques. From these methods, the adsorption method is the most widely used method to reduce the influence of environmental pollution [5]. This adsorption method has several advantages among others, the process is very simple, the costs used are quite cheap, and environmentally friendly [6].

The adsorption process is one of the waste treatment techniques that is expected to be used to reduce metal concentration or excessive organic composition [7]. The adsorbents commonly used in the adsorption process are activated carbon, silica gel, alumina, and zeolite. Recently, researchers have collected the processing of wastes contaminated with coloring agents and heavy metals using bacteria, fungi, yeast and algae [8]. Biomass is usually used as biosorbent because it is available in large quantities, mostly biochemically, very effective, and relatively stable. Absorption of substance molecules by algae biomass can occur through interaction with functional groups such as hydroxyl, carboxylic, amino and phosphate [9].

Recently it has been developed, the use of alternative adsorbents originating from nature because it is more economical, for example algal biomass. Some types of algae have received attention because of their high ability to adsorb ions or molecules through the existed functional groups in algal biomass and their use as biosorbents which can be reused for waste processing in solution. However, biomass also has several disadvantages, namely low density, easily damaged due to decomposition of other microorganisms, and difficult techniques to use in columns as adsorbents [10].

Therefore, it is necessary to make modifications to overcome the disadvantages of using algae as an adsorbent. One matrix that is often used in the processing of algal biomass is silica. Silica gel is an inorganic solid that has a surface active side such as silanol (-Si-OH) and siloxane (Si-O-Si) groups which can chemically bind to functional groups found in algae biomass and it has large surface area.

To further increase the adsorption ability of modified algae-silica biomass, silica-magnetite coating techniques were carried out [11]. Silica coating techniques with magnetite particles are used so that the adsorbent has a large capacity and selectivity for adsorbates adsorbed and separation of adsorbates can be used quickly [12]. Using this technique, it is expected that silica-*Spirulina* sp. algae hybrids with magnetite can produce adsorbents that are effective against MB and environmentally friendly because they have no by products such as suspended solids. In addition, using this technique cannot damage the structure so that the immobilization process of *Spirulina* sp. on silica matrix is expected to maintain the activity of functional groups provided in *Spirulina* sp. algae [13].

## 2. Experimental details

### *2.1. Shynthesis HASM*

Solution (A) 5 ml of TEOS mixed with 2.5 ml of distilled water added 0.1 g of magnetite put into a plastic tube, then added a few drops of HCl 1 M to pH 2. Solution (B) as much as 0.4 g of *Spirulina* sp. algae biomass mixed with 5 ml ethanol stirred with a magnetic stirrer until the solution is homogeneous. Both solutions are mixed until a gel is formed [14]. The wet gel formed is left to stand for 24 hours then washed with ethanol and distilled water until pH 7, dried using freeze dry, then crushed with a grinding tool [15].

### *2.2. Isotherm Adsorpsi*

An amount of 50 mg of HAS and HASM adsorbents with optimum concentration was put into 5 different plastic containers. Then added 20 ml of solution with different concentrations (0, 25, 30, 100, 200, and 300 mg $L^{-1}$) with optimum stirring time. The optimum pH was made using citrate buffer solution with pH range of 3.0 - 6.0 for the acidic pH and the phosphate buffer solution with pH range of 7.0-9.0 for the alkaline pH. After stirring, the adsorbent and solution were separated using centrifugation. The filtrate obtained was analyzed by UV-Vis spectroscopy at the maximum wavelength of MB 664 nm.

## 3. Results and discussion

### *3.1. Characterization Spektroskopi IR*

In Figure 1(a), it can be observed that the *IR* spectra of magnetite have an absorption band at wave number 547.78 $cm^{-1}$ which is a characteristic absorption of magnetic particles ($Fe_3O_4$). The absorption band in the wave number area 3441.01 $cm^{-1}$ originating from the hydroxyl group (-OH), at the wave number region 1627.920 $cm^{-1}$ is the carbonyl group. In Figure 1(b), it can be observed that *IR* spectra of *Spirulina* sp. algae biomass. There is an absorption peak of 3000 $cm^{-1}$ wavenumber which is a stretching region N-H which overlaps with O–H. In the absorption area of 2931.80 $cm^{-1}$ shows the stretching C-H vibration of (-$CH_2$) aliphatic and in the absorption area 1658.78 $cm^{-1}$ showed asymmetric carbonyl.

In Figure 1(c), it can be observed that the *IR* spectra of HAS show that the peak at wave number 1087.85 $cm^{-1}$ symmetrical vibration absorption of O atoms from the siloxane (Si-O–Si) functional group. The presence of a siloxane group reinforced by absorption at wave number 794.67 $cm^{-1}$ is a symmetrical stretching vibration of Si-O and a peak at 455.20 $cm^{-1}$ has a bending vibration of Si-O from a siloxane group (Si-O-Si). At wave number 3441.01 $cm^{-1}$, a wide band of sharp absorption peaks appears which is the -OH stretching vibration of a silanol (Si-OH) group. The absorption band area of 2931.80 $cm^{-1}$ shows stretching vibrations C-H of (-$CH_2$) aliphatic derived from organic groups in the biomass of *Spirulina* sp algae. This shows that the process of immobilizing algae biomass with silica as a matrix has been successfully carried out. In the absorption area 1658.78 $cm^{-1}$ showed asymmetric carbonyl.

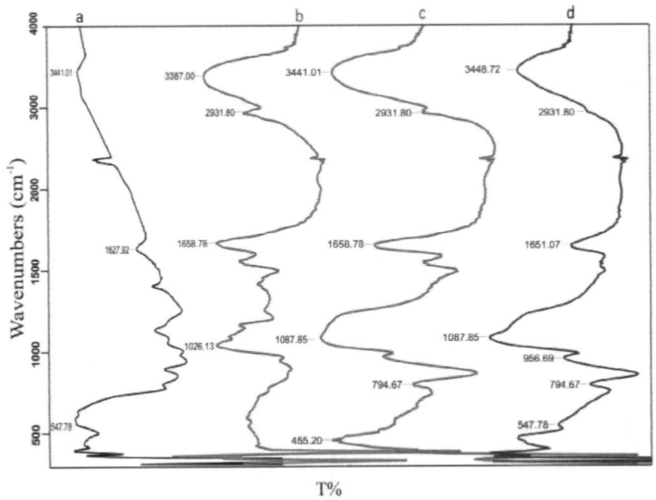


**Figure 1**. Spectra *IR* (a) Magnetite, (b) Biomassa algae *Spirulina* sp., (c) HAS, and (d) HASM

In Figure 1(d), it can be observed that the *IR* spectra of HASM, it is known that the absorption band at wave number 2931.80 $cm^{-1}$ is a stretch of C-H stretching vibration of ($-CH_2$) aliphatic originating from the organic group biomass *Spirulina* sp. This is not much different from the HAS spectra in the absorption indicating that the immobilization process of algal biomass with silica matrix and coating with magnetite particles has been successfully carried out. At wave number 547.78 $cm^{-1}$ which is a characteristic absorption of magnetic particles ($Fe_3O_4$). The absorption band at the wave number area 3448.72 $cm^{-1}$ originating from the hydroxyl group (-OH), in the wave number region 1651.07 $cm^{-1}$ is the carbonyl group.

### *3.2. Characterization SEM-EDX*

The differences in surface morphology is displayed in Figure 2 between (a) magnetite and (b) HASM, the magnetite shows the presence of granules of magnetite while in HAS it is more homogeneous because of the presence of silica, and HASM in the material appears homogen but there are still magnetite granules derived from $Fe_3O_4$. The EDX data in Figure 2(a) shows the presence of Fe and O elements derived from $Fe_3O_4$ and in Figure 2(b) it is known that the results of elemental analysis using EDX show the presence of C, O, Si, and Fe elements derived from *Spirulina* sp. algae biomass, matrix silica and $Fe_3O_4$. This supports the results of characterization using XRD, which indicates that the HASM material was successfully synthesized from HAS and magnetite.

IOP Conf. Series: Materials Science and Engineering **857** (2020) 012019 doi:10.1088/1757-899X/857/1/012019

**Figure 2**. Results of the 2000x SEM-EDX characterization (a) magnetite and (b) HASM

*3.3. Adsorption*

*3.3.1. Adsorption kinetics*

The adsorption rate of the adsorbent to MB acts to determine the optimum conditions of interaction of the MB solution and its equilibrium state to the adsorbent. In the plot of the amount of MB adsorbed against the interaction time of Figure 3(a) shows that the MB adsorption by HAS and HASM varies to reach the optimum time at each time of 15; 30; 60; 90; and 120 minutes. From Figure 3(b) shows that adsorption takes place quickly in the initial stage on the outer surface of the adsorbent followed by a slower internal diffusion process which is a step from determination of rate.

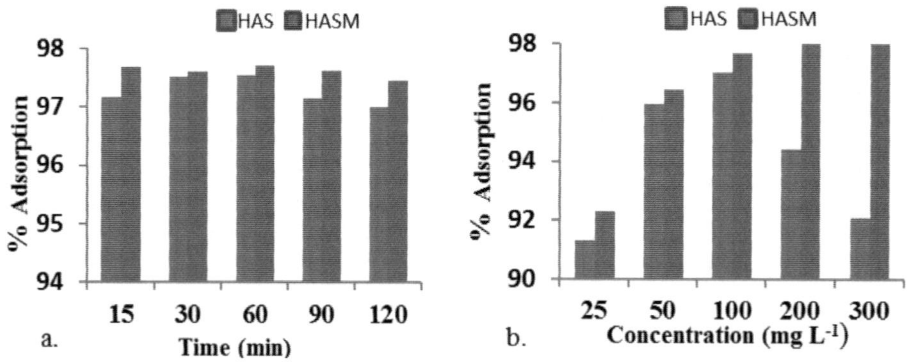


**Figure 3**. Influence of (a) interaction time and (b) concentration of MB adsorption by HAS and HASM.

The relatively short contact time can cause the adsorption process to be not optimal, this is because the adsorption speed is determined by the diffusion rate of the adsorbate species on the solids of the adsorbent. The longer the contact time of adsorption, the greater the collision frequency between the adsorbate particles and the absorbent, so that it can cause greater amounts of adsorbed MB [16]. Then from the data contained in Figure 4 was analyzed using pseudo first order and second order kinetics [11, 14].

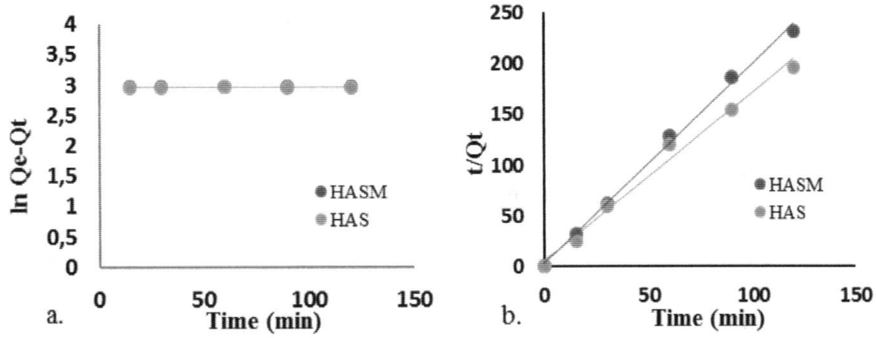


**Figure 4**. Kinetics of pseudo (a) first-order and (b) second-order MB dye on HAS and HASM

From Figure 4(a) it can be observed that the value of $R_2$ in the pseudo first-order kinetics equation for HAS and HASM is 0.307 and 0.308 respectively. This shows that the MB adsorption on HAS and HASM does not follow the pseudo first-order kinetics model. In addition to using pseudo first-order kinetics, analysis was also carried out using pseudo second-order kinetics. In Figure 4(b) it can be seen that the value of $R_2$ for each of the HAS and HASM adsorbents is 0.987 and 0.996, respectively. This indicates that the adsorption process in HAS and HASM tends to follow the pseudo 2 order kinetics model. In addition, from Figure 4 data shows that the $k_2$ value in the HASM material is greater than HAS. This indicates that HASM has a greater adsorption rate than HAS. The greater rate of adsorption rate on HASM compared to HAS was due to the magnetic properties of HASM which resulted in higher adsorption rates on HASM compared to HAS.

*3.3.2. Adsorption Isotherm model*

Concentration is one of the factors that determines optimum adsorption. In this study, a concentration variation of 0 - 300 ppm was used to determine the optimum concentration of MB absorbed. In Figure 3(b) it can be observed that adsorption increases with increasing concentration of MB. At the concentration of 300 ppm methylene blue, the amount of MB was adsorbed at HAS at 92.09% and at HASM at 98.03%.

The data contained in Figure 3(b) can be observed that the higher the initial concentration can be interacted with HAS and HASM adsorbents, the higher the amount of adsorbed metal (Q). Data were analyzed using the adsorption isotherm model. Adsorption isotherm is used to determine the interaction between the amount of adsorbed substance and the amount of adsorbent and the ability achieved by the adsorbent in adsorbing methylene blue. Adsorption isotherm is an important parameter in adsorption because it plays a role in determining the maximum conditions to produce optimal adsorption. In this study two patterns of adsorption isotherm equations were used, namely Langmuir and Freundlich isotherms on HAS and HASM.

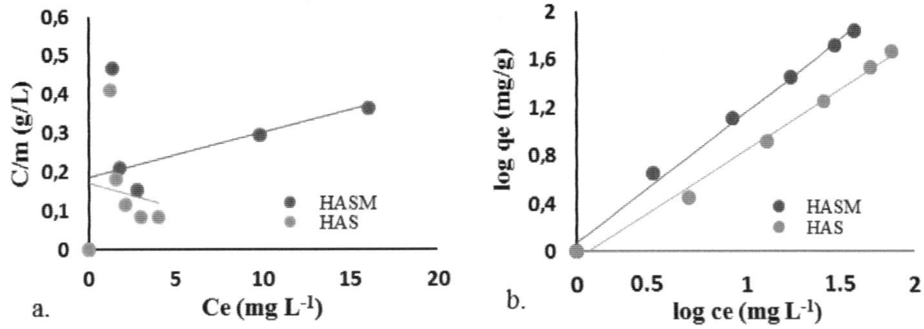


**Figure 5.** The linear equations of the adsorption isotherm models of the MB dye on the HAS and HASM adsorbent with the adsorption isotherm model of (a) Langmuir and (b) Freundlich

In the Langmuir isotherm it is explained that the adsorbent surfaces of HAS and HASM have active sites that are proportional to the absorbent surface area [17]. Each active site has the same energy so that it can be said that the surface of the adsorbent is homogeneous. In addition to being analyzed using the Langmuir isotherm model, analysis was also carried out using the Freundlich adsorption isotherm model. Freundlich adsorption isotherm states that the surface of the adsorbent is heterogeneous [18]. This is because the affinity of each active site is not the same, so the adsorption on the most active sites is preferred.

Based on the data presented in (Table 1) it can be observed that the pattern of MB adsorption isotherms in HAS and HASM tends to follow the Freundlich adsorption isotherm. This is indicated by the correlation coefficient ($R_2$) in the Freundlich isotherm parameters greater than the Langmuir isotherm parameters found in (Table 1). The adsorption capacity ($n_m$) in HAS and HASM for MB is 83.33 mg $g^{-1}$ and 90.90 mg $g^{-1}$.

**Table 1.** Langmuir and Freundlich adsorption isotherm parameters

| Adsorbent | Langmuir | | | Freundlich | | | |
|---|---|---|---|---|---|---|---|
| | $n_m$ (mg $g^{-1}$) | $K_L$ (L $mol^{-1}$) | $R^2$ | $n$ (mol $g^{-1}$) | $K_F$ (mg $g^{-1}$) (L $mg^{-1})^{1/n}$ | $R^2$ | $\Delta G°_{ads}$ (kJ $mol^{-1}$) |
| HAS | 83.33 | 0.069 | 0.015 | 1.098 | 1.153 | 0.993 | 6.668 |
| HASM | 90.90 | 0.058 | 0.193 | 1.101 | 1.153 | 0.994 | 7.101 |

The value of $K_L$ (adsorption equilibrium constant) on HAS for MB is 0.069 L $mol^{-1}$ and $K_L$ on HASM for MB is 0.058 L $mol^{-1}$. The greater the $K_L$ value, the greater the affinity of absorption by the adsorbent. The Langmuir isotherm explains that the absorbent surface in this case HAS and HASM there are a number of active sites that are proportional to the absorbent surface area. At each active site has the same energy so that it can be said that the surface of the adsorbent is homogeneous. In this study the MB adsorption process has adsorption energy of less than 20.92 k J $mol^{-1}$ so that it can be grouped to occur through physical interactions. However, the low adsorption energy does not indicate that adsorption only takes place in physics, but shows that chemical interactions play a role in adsorption in addition to physical interactions. The low adsorption energy can be caused by the binding mechanism of MB ions in HAS and HASM due to the large number of HAS and HASM functional groups that can bind MB ions and have different reactivity so that they can bind MB ions through physical and chemical interactions. Increased MB adsorption capacity in HASM is caused by reversible properties

## 4. Conclusions

The optimum amount of MB adsorbed by HAS and HASM adsorbents at pH 6 with a contact time of 60 minutes and tends to follow the pseudo-second-order kinetic model. Methylene blue adsorption isotherm by HAS and HASM tend to follow the Freundlich isotherm model with each adsorption capacity on HAS at 83.33 mg $g^{-1}$ while on HASM at 90.90 mg $g^{-1}$.

## Acknowledgements

The authors would like to thank the Directorate of Research and Community Service, Directorate General for Research and Development, Ministry of Research, Technology and Higher Education of the Republic of Indonesia who have funded this research in accordance with contract number: 065/SP2H/LT/DPRM/2019.

## References

[1] Buhani, Musrifatun, Pratama D S, Suharso, and Rinawati 2017 *Asian J. Chem.* **29** 2734-2739
[2] Seshadri S, Bishop P L, and Agha A M 1994 *Waste Manage.* **14** 127–137
[3] Sen S and Demirer G N 2003 *J. Environt. Eng.* **129** 595-601

[4] Buhani, Hariyanti F, Suharso, Rinawati, and Sumadi 2019 *Desalin Water Treat.* **142** 331–340
[5] Buhani, Suharso, and Sumadi 2010 *Desalination.* **259** 140-146
[6] Gupta S S and Bhattacharyya K G 2006 *J.Colloid. Interf. Sci.* **295** 21-32
[7] Buhani, Puspitarini M, Rahmawaty, Suharso, Rilyanti M, and Sumadi 2018 *Orient J. Chem.* **34** 2043–2050
[8] Gupta V K, Nayak A, and Agarwal S 2015 *Environ. Eng. Res.* **20** 1–18
[9] Celekli A, Atmaca, M Y, and Bozkurt H 2010 *J. Hazard. Mater.* **173** 123–129
[10] Buhani, Suharso, Luziana F, Rilyanti M, and Sumadi 2019 *Desalin Water Treat.* **171** 281–293
[11] Buhani, Rinawati, Suharso, Yuliasari D P, and Yuwono S D. 2017. *Desalin Water Treat.* **80** 203–213
[12] Peng Q, Liu Y, Zeng G, Xu W, Yang C, and Zhang J 2010 *J. Hazard.* Mater. **177** 676-682
[13] Liu Y, Zeng Y, Xu W, Yang C, and Zhang J 2010 *J. Hazard. Mater.* **177** 676-682
[14] Buhani, Suharso, and Satria H 2011 *Eur. J. Sci. Res.* **51(4)**: 467–476
[15] Entezari M H and Bastami T R 2013 *Mater. Res. Bull.* **48** 3149–3156
[16] Idris M N and Ahmad Z A 2011. *Int. J. Sci. Basic Appl. Res.* **11** 305-311
[17] Buhani, Herasari D, Suharso, and Yuwono S.D 2017 *Orient. J. Chem.* **33 (1)**: 418-429
[18] Buhani, Suharso, and Fitriyani AY 2013. *Asian J. Chem.* **25(5)**:2875–2880

# Denoising of partial discharge waveforms using multivariate wavelet method

**H H Sinaga[1,*], H B H Sitorus[1], D Permata[1], Y Yuniati[1], R B Vraja[1]**

[1] Department of Electrical Engineering, Faculty of Engineering, Universitas Lampung, Jalan Prof. Soemantri Brojonegoro No. 1, Bandar Lampugn, 35143 Indonesia

*Email: herman.h.sinaga@eng.unila.ac.id

**Abstract.** The presence of Partial discharge (PD) in high voltage equipment is an indication of insulation defect or degradation in its insulation. The PD presence can be detected using an electromagnetic sensor, such as fractal sensor. The signals picked-up by the sensor then recorded using a digitizer or an oscilloscope. During the signals pick-up proses, the sensor will capture all the electromagnetic signals around the sensor. Thus, the sensor not only pick-up the PD signals but also unwanted signals or noises. The PD signals captured by the sensor might be buried in heavy noise that the PD signals might be unidentified. To recognize the PD signals de-noising process can be done. In this paper, discussed the denoising process by applying multivariate wavelet to the PD signals captured by using fractal sensor. Results show de-noising process can eliminated noises from PD waveforms even if the PD signals buried by noise signals.

**Keywords:** Partial discharge, fractal sensor, multivariate wavelet, insulation

## 1. Introduction

Partial discharge is a symptom of the degradation of the transformer insulation. If the partial discharge allowed to presence in long period of time thus grow overtime, the partial discharge can lead to the insulation breakdown and might cause transformer failure. Thus, partial discharge detection on transformer is a very essential.

Partial discharge detection on transformer can be done using varying method. Conventional method as defined by IEC standard 60270 is the most popular method used to detect PD in transformer. This method has mayor advantage compare to other method i.e. its ability to calibrate the measurement results. The IEC standard 60270 method results can convert to amount of pC generated by the PD sources. However, this method has a major drawback when applied in open area with high surround activity. As the conventional method works in frequency range up to 1 MHz [1], the surround electric field can interfere the measurement results. Thus, conventional is very risky to the noise signals around the transformer being test.

Partial discharge detection aforementioned in IEC standard 60270 classified as conventional method and other methods named as non-conventional. Non-conventional method such as DGA-dissolve gas analysis has proven accurate to detect the presence of PD inside the transformer tank. However, DGA has substantial disadvantage i.e. partial discharge can be detected only after some gas released by the PD source has built-up in the liquid isolation. The gas built-up, however, can take quite long time.

One of PD detection method developed lately is by detection the electromagnetic signals generated by the PD source. As the source to detect the PD events is electromagnetic signals generated by the PD events, this method is also called as ultra-high frequency (UHF) method. Though detection also can be arranged for lower frequency [2], such as 50 MHz ~200 MHz. Higher frequency used in UHF method provide advantage compare to a lower frequency one. This due to noise disruption relative lower. However, though electromagnetic PD detection has lower noise interferences the noise might still disrupt the PD signals thus error in PD detection can occur.

## 2. Detection of Partial Discharge Electromagnetic Signal

In the event of a discharge in transformer insulation, free electron charges which were initially at rest are accelerated and decelerated by an external force. The acceleration and deceleration processes produce a time varying electromagnetic field which radiates outward from the PD location.

A stationary charge has only electric field which is radiated radially. As the charge is in motion, electric field and magnetic field are produced. Assume the charge is initially at rest then accelerated. The electric field lines at any point in the direction of the path of the moving charge are entirely radial and must be continuous since they are produced by the same charge.

During the acceleration of the charge, the electric field lines are always updating its position. However, due to the time needed by the electric field lines to adjust, the lines will have disrupted the direction or become misaligned. This line disruption is known as a 'kink' which have both a static Coulomb field and an electric field which are perpendicular to each other. The transverse electric field produced by this process then causes radiation. The maximum radiation occurs along the line perpendicular to the direction of the acceleration. In the same direction with the acceleration there is no transverse electric field component produced, only a static Coulomb field thus no radiation occurs.

The basic diagram of the UHF PD detection method is shown in Figure 1. The main component to detect the PD signals is an antenna which acts as a sensor that picks up electromagnetic signals emitted by the PD sources. The sensor is connected to a measurement unit to show and record the PD signals. In cases when the PD signal is too small, an amplifier can be installed between the sensor and the measurement unit. Using an amplifier which has a specific operational frequency, the PD signals can be magnified whilst blocking the noise signals. In the UHF range, the noise mainly comes from known communication sources such as digital TV or mobile phone signals. Thus, using a specific amplifier which excludes known noise frequencies, a clear PD signal can be shown and recorded by the measurement unit.

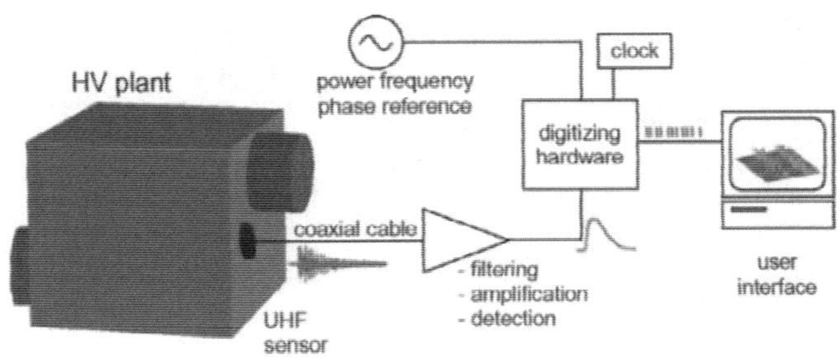


**Figure 1:** Typical UHF PD detection diagram [3].

### 2.1. *Sensor to detect Partial Discharge on Transformer*

The purpose of transformer monitoring extends from PD detection and recognition to PD localization. To be able to undertake UHF PD transformer monitoring, a sensor with capabilities to detect signals in the UHF frequency range is needed. For power transformer monitoring, the UHF sensor is inserted into the transformer tank to capture the electromagnetic waves emitted by the PD source. There are two ways of installing the sensor: via the oil drain valve [4] or the dielectric window [5]. The size of the oil drain valve imposes a constraint on the sensor dimension, while the dielectric window can be created with an appropriate size to accommodate the sensor. However, the placement of a dielectric window sensor needs an additional hole to be fabricated on the transformer tank. As for the oil valve sensor, this is not required because the sensor can be easily retrofitted into the transformer via the existing built-in oil drain valve [4].

Typically, the sensor is a monopole type. It can be inserted via the oil drain valve. The size of the sensor is usually limited by the diameter and length of the oil drain. A typical oil drain valve sensor size is 5 cm in diameter, and 10 cm in length [4]. The shape of the sensor can be a short monopole [4, 6], plate, zigzag or conical [4, 7] or any shape as long as it is able to be fitted to the drain valve. The sensitivity of this kind of sensor is affected by the depth of the sensor insertion [4]. The deeper the sensor is inserted, the higher the magnitude of the PD signals acquired. However, the sensor must not initiate breakdown due to the high electric stress at the tip of the sensor [5]. To reduce electric stress, the sensor can be encapsulated in some dielectric material [5].

For a dielectric window, the sensors usually have a planar shape [5]. The sensor can be a micro-strip sensor [8,9,10], log-spiral, spiral [5,11] or fractal [12]. This kind of sensor is usually etched on the surface of a dielectric material, using the same process as in making electronic printed circuit boards (PCB). The sensor is etched on the PCB with dimensions proportional to the working frequency of the sensor. In 37, the miniaturization of the microstrip UHF antenna was discussed by applying an impedance matching technique. This technique can reduce the antenna dimensions to 5x5 cm for example. However, even though the antenna works in UHF range, the bandwidth is very limited. To increase the bandwidth, [10] designed a microstrip antenna using sandwich substrate. As a result, the frequency range of the antenna has a very wide range from 30 MHz to 3000 MHz. However, to manufacture this microstrip antenna is not practical.

Most of the measuring arrangements are unbalanced systems, i.e. the input to the measuring system is a coaxial cable which consists of a live input and ground. While planar sensors are usually crafted as a balanced system, an unbalanced system is also possible, such as a circular or single arm log-spiral or spiral. Thus, to connect the sensor to the measuring system, a converter from balanced to unbalanced is needed. This connector forms part of the sensor and is called a balun.

In this paper, we use fractal sensor to detect the PD signals. As the sensor is an un-balance sensor type, the sensor connected directly to an oscilloscope via coaxial cable.

**Figure 2**. Fractal sensor in used in the experiment.

### 2.2. *Fractal Sensor*

In this paper discussed the using of fractal sensor to capture the partial discharge signals. The sensor has dimension of 5x5 cm. as shown in figure 2. The sensor built on single layer PCB with terminal connected to a BNC. The sensor VSWR is shown in Figure 3. As one can see the sensor VSWR shows good value at the designed frequency range as the VSWR value below 2 for the frequency up to 1000 MHz.

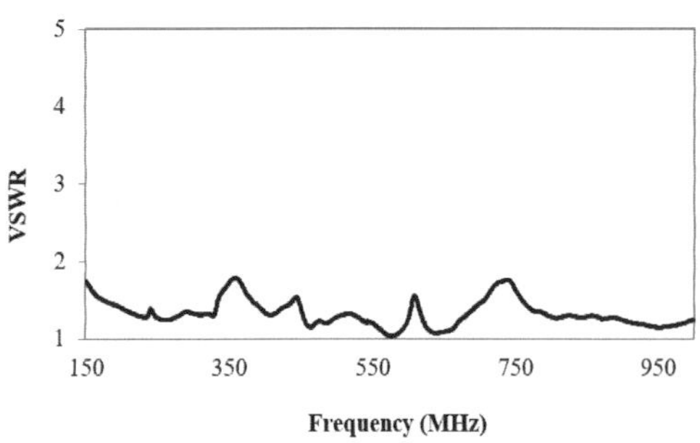


**Figure 3.** Fractal sensor VSWR up to 1000 MHz.

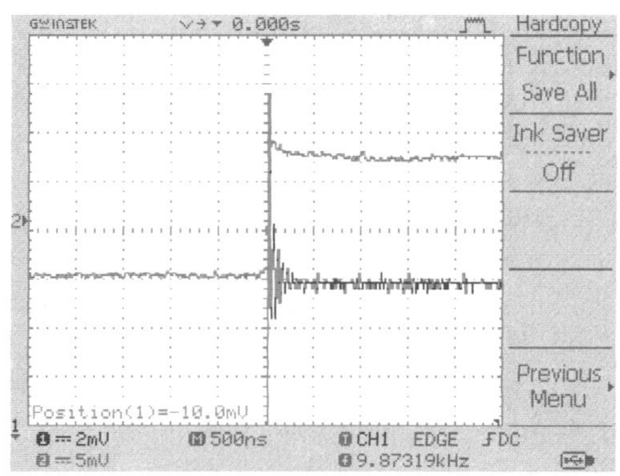


**Figure 4**: Sensor response to a step pulse.

### 2.3. *Sensor Response to a Step Pulse*

Partial discharge signals is a very short pulse that contain frequency up to a few GHz [13]. Thus, to test the sensor ability to detect the PD signal one can use a steep pulse which contain frequency in a wide range up to GHz. The step pulse response of the fractal sensor which use in the experiment shows in figure 4. The sensor shows ability to capture the step pulse signals generated in an TEM cell. So, the fractal sensor can be used to detect the partial discharge signals in the experiment.

## 3. Noise on Partial Discharge Signals

On-site partial discharge detection might encounter a very different scheme. The partial discharge detection on laboratory, within controlled environment, will encounter less noise or even eliminated. However, on-site partial discharge detection might encounter noise from the surrounding environment. Noise can be produced by varying sources and can be classified in two type i.e.: 1. continue noise produce by telecommunication system, thyristor switching and line transmission switching, and 2. thermal noise produce by the PD detection system itself. The noise can be swamped the PD signals, so the PD signals unable to recognize. To recognize the PD signals, denoising might needed.

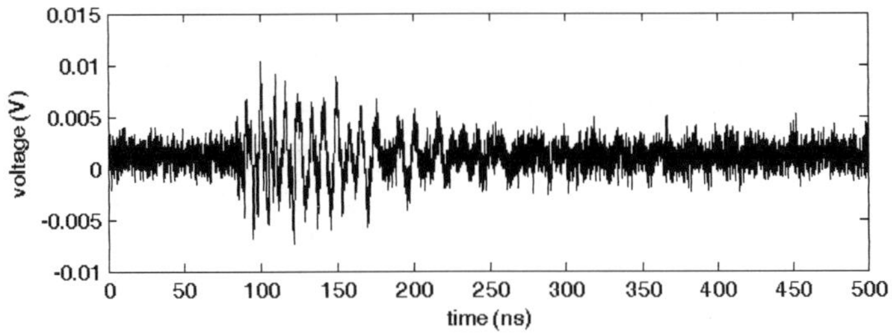


(a)

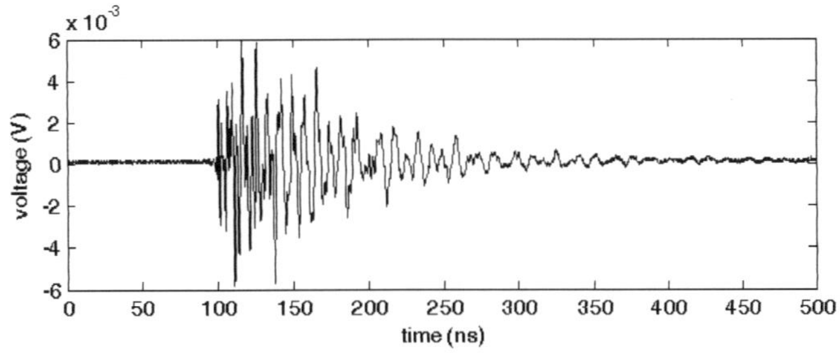


(b)

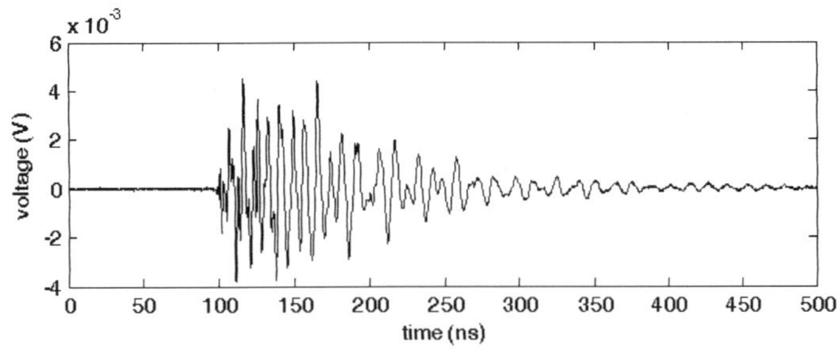


(c)

**Figure 5 :** Typical denoising result using multivariate wavelet method (a) original waveform, (b) denoising using threshold value, and (c) final result after applying PCA component analysis [15].

### 3.1. *Denoising process*

A free noise of partial discharge waveforms can be achieved by applied denoising process to a noisy partial discharge waveform. In this paper, discussed the using of wavelet multivariate [14] to denoise the partial discharge waveforms. The method based on regression model:

IOP Conf. Series: Materials Science and Engineering **857** (2020) 012020 doi:10.1088/1757-899X/857/1/012020

$$X(t)=f(t) + \varepsilon(t),\ t = 1,\dots,n. \quad (1)$$

where:

$(X(t))_{1\leq t\leq n}$ = observed waveforms

$f$ = denoised waveforms

$\varepsilon(t)_{1\leq t\leq n}$ = Gaussian white noise with unknown variance value $\sigma^2$

The multivariate denoising procedure is done in two steps. First, the original waveform is arranged by using wavelet transform, and then by using applying multivariate thresholding the waveform is denoised. Second, principal component analysis (PCA) is used to improve the denoising results. Figure 5 [15] shows typical wavelet multivariate denoising results in two steps procedures.

## 4. Experiment Setup

Figure 6 shows the experiment diagram. The partial discharge source is a surface discharge type. The sensor installed near the PD source i.e. at distance around 20 cm. The sensor connected to an oscilloscope to record the PD waveforms. The recorded data than copied to a PC, for further analysis. Input voltage controlled using a voltage regulator. The input voltage increased unti the PD generated by the PD source.

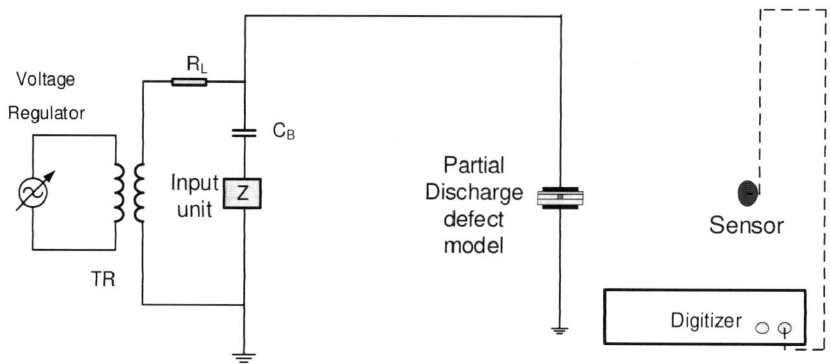


**Figure 6.** Diagram of the experiment set-up

## 5. Results and Discussion

The denoising process is done using waveforms produce by a surface discharge source. The PD signals captured by a fractal sensor recorded by using a digitizer or oscilloscope. Two waveforms are use as input of the denoising process. The first waveform has amplitude at around ~1 mV and the second one has amplitudes ~0.8 mV. A white-noise is used as to alter the original waveforms, with amplitude of 0.5 mV and 0.75 mV respectively, named as case I dan case II.

Figure 7, 8 and 9 show the denoising process for the two partial discharge waveforms. Figure 7 shows the white-noise with amplitude 0.5 mV used to flooded the partial discharge waveform. Figure 8.a and 9.a are two original partial discharge waveforms. The two original waveforms than swamped with the white-noise to produce two noisy waveforms, as shown in figure 8.b and 9.b. The multivariate wavelet than applied to denoise the noisy waveforms. Figure 8.c and 9.c show the denoised waveforms after applied multivariate wavelet denoising method.

By comparing the original waveforms with the denoised waveforms, it can be seen the two denoised waveforms have similar characteristics with the original waveforms. The magnitude of the original and the denoised waveforms have almost similar magnitude. The denoising process able to reconstruct the waveforms although buried by high noise as the original and the denoised waveforms shows similar forms. However, some of the added white-noise still recognizable in the denoised waveforms. This can be seen in front of the denoised waveforms.

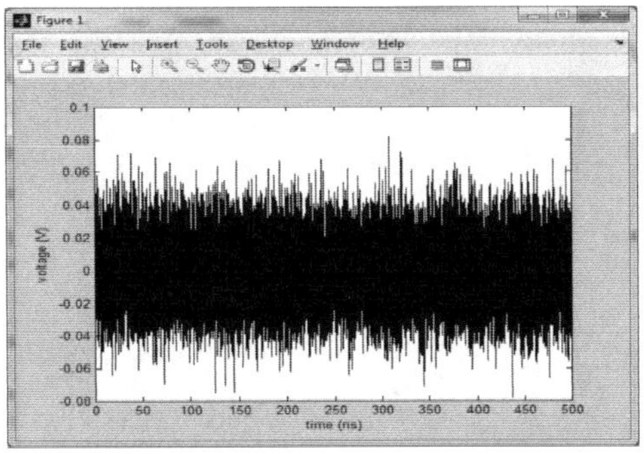


**Figure 7.** Gaussian white noise

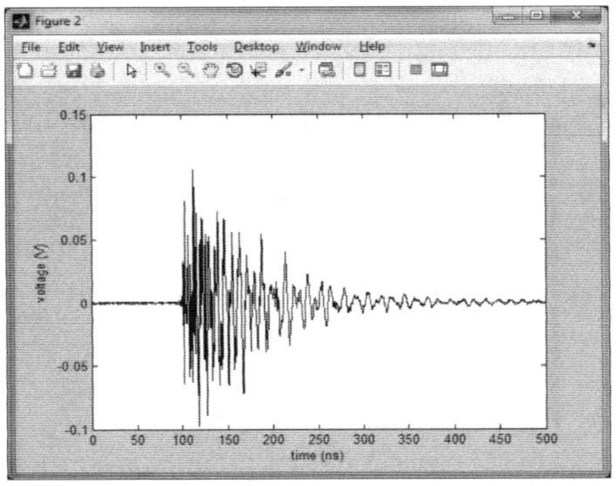


(a)

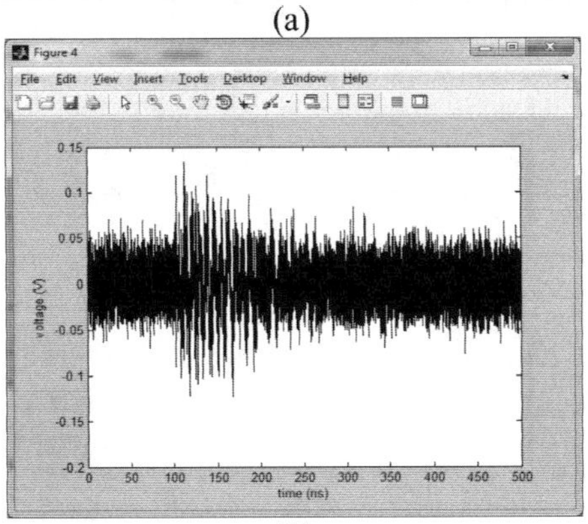


(b)

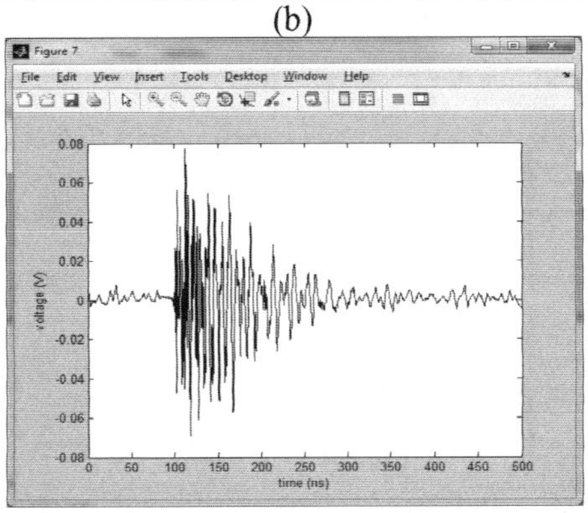


(c)

**Figure 8**. The denoising process of waveform I, a. Original waveform, b. white noise added to original waveform, and c. the denoised

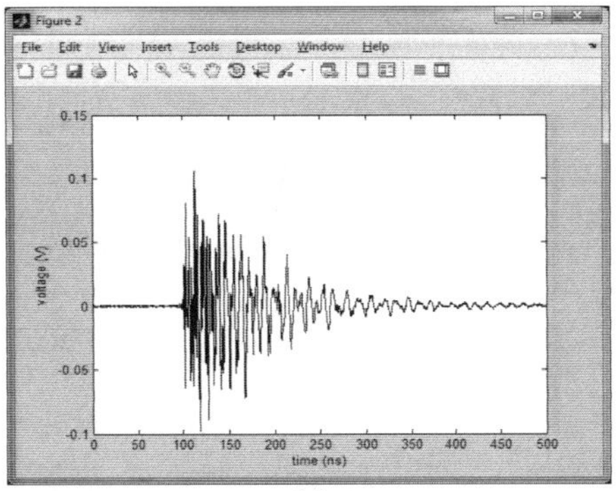


(a)

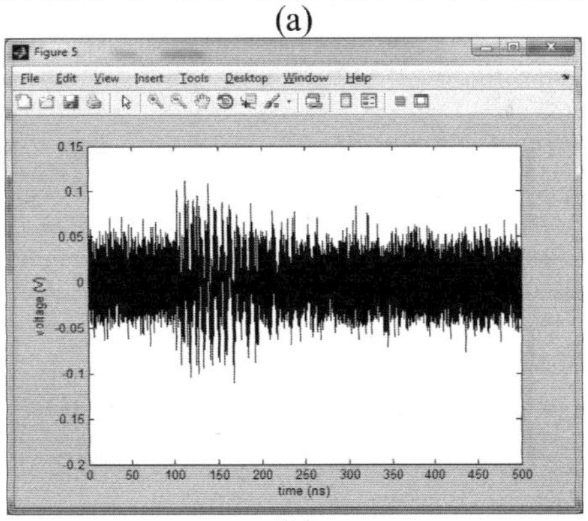


(b)

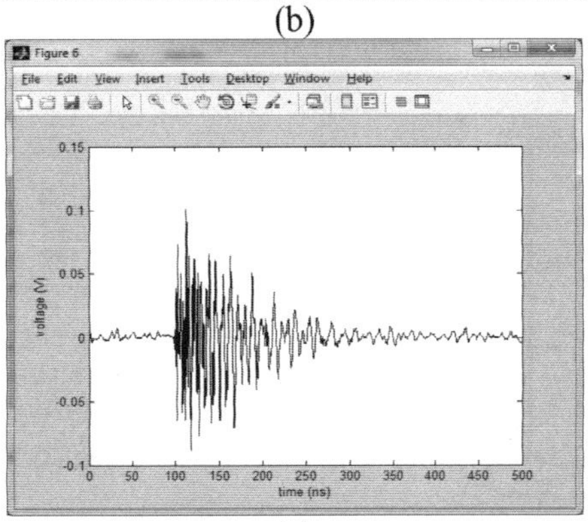


(c)

**Figure 9.** The denoising process of waveform II, a. Original waveform, b. white noise added waveform using multivariate wavelet method.

to original waveform, and c. the denoised waveform using multivariate wavelet method.

Effectiveness of the denoising method is determine as below:

1. Signal to noise (SNR) between partial discharge waveform with the white-noise is calculated using:

$$SNR1_{db} = 10\log\left(\frac{P_{signal}}{P_{noise}}\right) \quad (2)$$

Where $P_{noise}$ is the white-noise and $P_{signal}$ are the original partial discharge waveforms.

2. The white-noise and the partial discharge waveform then summed-up. The summed-up waveforms then denoised using multivariate wavelets. The new SNR then calculated as:

$$SNR2_{db} = 10\log\left(\frac{P_{signal}}{P_{denoise}}\right) \quad (3)$$

Where $P_{signal}$ is the denoised waveforms and $P_{denoise}$ is the subtraction of the denoised signal with the original waveforms.

The calculated SNR of the denoising process is shown in Table 1. One can see, that the denoising process resulted a very clear waveforms with SNR2 value above 10 dB. Thus, it can be said the denoised waveforms is already clear from the noise.

**Table 1.** Signal to noise (SNR) ratio denoised waveforms

| Case | Waveform | SNR1 (dB) | SNR2 (dB) |
|---|---|---|---|
| I | 1 | -1.8673 dB | 15.2147 |
| | 2 | -3.7000 dB | 13.5063 |
| II | 1 | -5.4087 | 13.6066 |
| | 2 | -7.2414 | 11.8632 |

## 6. Conclusion

Multivariate wavelet denoising method shows ability to incite the original waveform even when the denoised waveform buried in heavy noise. The denoised waveform has characteristic similar to the original waveforms. The denoised waveforms have a good result as shown y the high SNR value. In this research the SNR of the denoised waveforms to the noise signal is greater than 10 dB.

**Acknowledgements**

The authors would like to express their gratitude to Kemenristek-DIKTI Indonesia for supporting this research. This research was supported by Kemenristek-DIKTI Indonesia with the Fundamental Research Grant, contract number: 857/UN26.21/PN/2019.

**References**

[1] Bartnikas R 2002 *IEEE Transaction on Dielectric and Electrical Insulation* **9** pp 763-808
[2] Pinpart T and Judd MD 2009 *Electrical Insulation Conference (EIC 2009)* pp 26- 30
[3] Judd M D, Cleary G P, Bennoch C J, Pearson J S and Breckenridge T 2002 Electrical Insulation *Conference Record of the IEEE International Symposium on Electrical Insulation* pp 145-149.
[4] Roldan J L, Tang T and Gaskin M 2008 *IEEE Transaction on Dielectrics and Electrical Insulation* **15** 6 pp 1634- 1639
[5] Warne D and Haddad A 2004 *Advance in High voltage Engineering* (London - Institution of Electrical Engineers)
[6] Pinpart T. and Judd M D 2009 *Electrical Insulation Conference (EIC 2009)* pp 26- 30

[7] Agoris P, Meijer S and Smit J J 2007 *The Powertech '07 Conference* pp 2065 – 2069
[8] Ando A, Kagoshima K, Kondo A and Kubota S 2008 *IEEE Trans. On Antennas and Propagation* **56** 8 pp 2747-2751
[9] Marrocco G 2008 *IEEE Antennas and Propagation Magazine* **50** 1 pp 66-79
[10] Erentok A and Ziolkowski R W 2008 *IEEE Trans. On Antennas and Propagation* **56** 3 pp 691-707
[11] Roy A, Ghosh S and Chakrabarty A 2007 *Intl. Conf. on Industrial and Information Systems (ICIIS 2007)* pp 181-185
[12] Thaysen J, Jakobsen K B and Appel-Hansen J 2000 *Applied Microwave and Wireless* **12** 10 pp 40-50
[13] Judd M D and Farish O, *The International Symposium on Electrical Insulation (ISEI)* **2** pp 436-439
[14] Aminghafari M, Cheze N and Poggi J M *Computational Statistics & Data Analysis* **50** pp 2381-2398
[15] Sinaga H H, Phung B T and Blackburn T R 2014 *IET Generation, Transmission & Distribution* **8** 1pp 160 – 169

# Delignification of Abaca Fiber (*Musa textilis*) as Potential Substitute for *Eucalyptus pellita*

**A Darmawan[1], B Irawan[1], H Ni'mah[1,*], A Roesyadi[1], F Kurniawansyah[1]**

[1] Department of Chemical Engineering, Faculty of Industrial Technology, Institut Teknologi Sepuluh Nopember (ITS), Surabaya, East Java, Indonesia, 60111.

*Email: hikmatunn@gmail.com

**Abstract.** Five years old *Eucalyptus pellita* is a raw material used to manufacture pulp in Indonesia. In the field, the cropping cycle of this wood is still considered too long, so the need to develop other alternative raw materials that have characteristics similar to the fiber planting a shorter cycle. One of the alternative raw materials is abaca fiber (*Musa textilis*). Abaca fiber is preferred over other fibers because of its good tensile strength and resistance to rotting. In the present work, delignification of abaca fiber was carried out by kraft pulping process and characterized in term of their mechanical properties. Abaca fibers were treated with kraft pulping and oxygen delignification. Eucalyptus kraft pulp was also characterized as a comparison. Kappa number and viscosity decreased when the active alkali charge increased. Kraft pulp abaca with high cooking results and good viscosity is easily delignified by converting oxygen to low kappa numbers (7-9) without significant loss in viscosity. In addition, from the final results, the kappa number and viscosity, it was shown that abaca pulp has very high strength. Although oxygen delignification for lower kappa quantities can be considered, in terms of the properties of the pulp the abaca fiber is suitable as a raw material for pulp.

***Keywords*:** abaca fiber, kraft pulping, oxygen delignification

## 1. Introduction

As the pulp and paper industry grows more rapidly, the demand for pulp raw materials also increases. Pulp raw materials are generally derived from broad wood plants and needle leaf wood plants. At present, 5-year-old *Eucalyptus pellita* and *Acacia mangium* are the raw materials used for pulp production in Indonesia. The problem in the field is that the wood planting cycle is still considered too long so that it needs to develop other alternative raw materials that have fiber characteristics similar to shorter planting cycles [1]. One alternative raw material that can be used is abaca banana fiber (*Musa textilis*). Abaca fiber (*Musa textilis*) is a non-wood fiber (non-wood fiber) that can be used as an alternative material in making pulp because it contains about 68.32% cellulose, 19% hemicellulose, and 12-13% lignin [2]. Abaka is a superior fiber with high tensile and folding strength, high porosity, resistance to saltwater damage, and a long fiber length of up to 3m.

Abaca (*Musa textilis* Nee) is a type of banana plant which produces fiber, including in the family Musaceae. Abaca is known by several names such as banana fiber, Manila Hemp, Manila Henep, kofo tree or hote. In some areas in Indonesia, abaca plants are known by various names such as Banana

Manila (Manado), Cau Manila (Sunda), Kofo sangi (Minahasa). The main production of this abaca plant is abaca fiber which was originally only used as a raw material for making ship ropes and clothing materials [3]. However, further research along with the development of technology and human civilization has been able to maximize the use of abaca fibers. Abaca fibers are known as high-quality fibers that are used as raw materials for various industries such as textiles, paper and even automotive [4]. Abaca fiber has been used to make submarine cable wrappers, rigging on ships, paper money, checks, filter paper and wrapping paper (Bank Indonesia). In the automotive field, Mercedes Benz has used a mixture of thermoplastic polypropylene and abaca threads on car body parts [2].

For abaca, in some studies still using the soda pulping process [1] [2]. This method is very wasteful with chemicals, especially soda. Kraft process is known to produce pulp with a higher strength compared to the soda process, in this process all types of wood can be used and the benefits gained from this process are the process of recovering relatively high chemicals and producing energy that can be reused. Kraft process is a type of pulp manufacturing process that is commonly used by the pulp industry in Indonesia with raw materials, especially the type of broadleaf wood (hardwood). The most important factor in the kraft process is the concentration of the cooking solution. The increase in concentration causes an increase in the rate of delignification and results in an increased effect of cellulose solution. Active alkali (AA) concentration on the weight of wood that is often used is 15-20%. An increase in active alkali concentration that is too high causes greater cellulose damage than lignin, thereby reducing the yield and strength of the pulp.

Depending on the order of subsequent bleaching and the target paper level, kraft pulp that is not bleached with kappa number 18-30 is usually made from eucalyptus [3–5]. However, the element chlorine is still commonly used in pulp bleaching. Several studies have also reported oxygen delignification of the eucalyptus kraft pulp. However, there are no studies that explicitly aim at achieving very low kappa numbers for abaca pulp. An effective reduction in the content of lignin residues in kraft pulp without a negative effect on the strength properties of the pulp will be very important. on the one hand, for environmental protection and, on the other hand, to reduce bleaching costs.

The aim of this study is to investigate the possibility of producing abaca pulp with a low residual lignin content for bleaching using conventional kraft pulp followed by oxygen delignification. This research is part of a larger research project that aims, in addition to cooking experiments with non-wood raw materials, on the detailed characterization of liquids released related to cooking, oxygen delignification. so that it has the potential to reduce the use of wood as pulp raw material.

## 2. Materials and Methods

### *2.1 Materials*

The materials used are abaca fiber (*Musa textilis*) from Aceh and *Eucalyptus pellita* wood from Jambi Province (Table 1). The chips and fiber are filtered (SCAN-CM 40:94), dried and stored at 25% water content. The concentration standard of NaOH and Na2S for pulping was set to 97.0 g / l and 32.4 g / l respectively.

**Table 1.** Comparison of Chemical Properties of Abaca Fiber With Chip *Eucalyptus pellita* [6][7][8][9]

| Wood and Fiber Characteristics | Abaca Fiber | Chip *Eucalyptus Pellita* |
|---|---|---|
| Density (g/cm$^3$) | 1.5 | 0.469 |
| Cellulose | 68.32% | 68.85% |
| Hemi cellulose | 19% | 18% |
| Lignin | 12-13% | 28.09% |
| Moisture content | 10-11% | 31.6% |
| Ash content | 4.8% | 0.4% |

*2.2 Kraft pulping*

Cooking liquor including sodium hydroxide and sodium sulfide were prepared from Merck products. Concentrations of cooking liquor were analyzed according to SCAN–N 2:88 standard. Then, on the basis of concentration of cooking liquor and the ratio of liquor to chips weight (L/W). The cooking was carried out in a rotating digester *Haato 6 chambers* with active alkali charge of pulping liquors. The cooking conditions are shown in Table 2. Each chamber of digester was filled with 300 gram of oven-dried chips and the wood ratio between chips and liquid was 1: 3,5. The chips, white liquor, and solvent were added into chambers and placed in digester. Impregnation time consisted of 3 stages, the first stage was from 27 to 100°C for 30 minutes, the second stage was from 100 to 132°C for 30 minutes and the third stage was from 132 to 165°C for 30 minutes. The cooking process was at a maximum temperature of 165°C for 90 minutes. The resulted pulp was sieved through a screen somerville. Pulping yield was obtained gravimetrically by weighing the oven-dry pulp and reported as percentage of the original wood. Viscosity and kappa number were determined by TAPPI T 236 cm-85 and TAPPI T-230 om-89 respectively.

**Table 2.** *Cooking Conditions*

| Parameters | Conditions |
|---|---|
| Active Alkali (%) | 16, 17, 18, 19, and 20 |
| Sulfidity (%) | 25 - 28 |
| Maximum temperature (°C) | 165 |
| H-Factor | 800 |
| Wood chips dry mass (g) | 300 |
| Liquid ratio | 3.5 |

*2.3 Oxygen delignification*

Oxygen delignification is carried out in a rotating reactor with an electric heater and a volume of 20 liters, consisting of 6 individual tubes of 100 mL each. Pulp samples with 4 grams each were placed into the reactor in a medium consistency in Table 3 and heated to a predetermined temperature. The pulp is washed with distilled water. Pulp that has been washed, then filtered, and dried then tested its quality. The experimental design used is completely random in factorial settings.

**Table 3.** Oxygen Delignification Conditions

| Parameters | Conditions |
|---|---|
| Pulp consistency (%) | 10 |
| Reaction time (min) | 60 |
| Maximum temperature (°C) | 85-90 |
| Alkaline charge (% NaOH) | 2 |
| Wood chips dry mass (g) | 4 |
| Oxygen preasure (bar) | 20 |

*2.4 Delignified Pulps Analysis*

Kappa number of pulp and pulp viscosity were determined in accordance with the procedures of TAPPI T 236 cm-85 and TAPPI T-230 om-89 respectively. Lignin content, Delignification degree (DD), and Degree of Polymerization (DP) of pulping was calculated following the formula as follows:

$$Lignin\ Content\ (Klason) = 0.152 \times Kappa\ Number \quad (1)$$

$$DD\ (\%) = \frac{total\ lignin\ of\ wood - residual\ lignin\ of\ pulp}{total\ lignin\ of\ wood} \times 100\% \quad (2)$$

$$DP^{0.905} = 0.75\ [\eta]$$
$$[\eta] = instrinsic\ viscosity \quad (3)$$

## 3. Result and Discussion

### *3.1 Kraft pulping*

The kraft process is carried out to determine the pulp properties of abaca. The kraft method is 25% sulfidity, 16-20% active alkali (AA) at a cooking temperature of 165°C and a cooking time of 90 minutes at maximum temperature, as is commonly done in the pulp and paper industry. The main purpose of this cooking process is to reduce the lignin content measured from the kappa number (KaNo) according to the minimum standard as a reject pulp. The nature of the pulp produced from the process is presented in Table 4. The increase in active alkali load causes the total pulp yield to decrease as shown in Table 4. The higher the active alkali load used during the cooking process, the lower the pulp yield. This is caused by increasing levels of delignification. The amount of delegated lignin results in better physical quality of the pulp. However, cooking with an excessive active alkali load can also cause extensive fiber degradation [10]. The use of an active alkali load that is too low will also produce more unexpected rejection in the pulp. Reject is caused by wood chips and raw fiber in the cooking process because once the delignification process is not completed, the pulp quality decreases. Therefore, the use of an active alkali load needs to be considered to get the best pulp quality.

**Table 4.** The pulp properties of Abaca and *Eucalyptus pellita* (cooking conditions, see Table 2)

| Material | Active Alkali (%) | Total Yield (%) | Screen Yield (%) | Kappa Number | Viscosity (ml/g) | Lignin Klason (%) |
|---|---|---|---|---|---|---|
| *Eucalyptus pellita* | 16 | 53.14 | 52.84 | 16.69 | 954 | 2.53688 |
| | 17 | 53.2 | 53 | 15.87 | 918 | 2.41224 |
| | 18 | 51.92 | 51.82 | 14.99 | 892 | 2.27848 |
| | 19 | 50.77 | 50.7 | 14.6 | 847 | 2.2192 |
| | 20 | 50.6 | 50.6 | 13.4 | 810 | 2.0368 |
| *Musa textilis* | 16 | 58.6 | 58.4 | 13.24 | 955 | 2.01248 |
| | 17 | 55.44 | 55 | 12.45 | 939 | 1.8924 |
| | 18 | 52.54 | 52.32 | 12.2 | 917 | 1.8544 |
| | 19 | 50.35 | 50.22 | 10.85 | 880 | 1.6492 |
| | 20 | 48.2 | 48 | 10.15 | 848 | 1.5428 |

The increase in active alkali load affects the decrease in pulp yield. It can be concluded that every 1% increase in active alkali content will reduce 0.15% pulp yield. The high decrease in hardwood yield due to high hemicellulose content [6]. For *Eucalyptus pellita*, the maximum pulp yield is 17% active alkali. Carbohydrates, cellulose, and hemicellulose are very unstable in alkaline solutions. During the process of cooking with alkalis, damage to glycosidic bonds occurs. This results in the separation of the main chain chains and carbohydrate chains. Termination produces low molecular weight soluble compounds which cause reduced yield. Termination of carbohydrate chemicals during the process of making alkalis involves several types of reactions that are influenced by concentration but do not depend on the presence and ions used in the kraft process.

The high kappa number shows the high lignin content in the pulp produced after the cooking process. Pieces of raw wood and fiber contain cellulose, hemicellulose and other impurities. Active alkali charge was observed to get the target kappa number. Within the normal range, the kappa number is expected to be <20 according to the alkaline preparation standard where the range of the kappa standard for *Eucalyptus pellita* is 12-18 [8]. A decrease in kappa number is caused by the use of an excessive active alkaline load which damages cellulose fibers and not only lignin is dissolved but also cellulose is dissolved in liquid. Figure. 1 shows that the kappa number decreases with an increasing active alkali load. This is due to the increased degradation of lignin with a higher active alkali load in the liquor used. At the active alkaline load level of 16-20% for *Musa textilis* and *Eucalyptus Pellita*, the standard kappa number ≤18 is reached. However, *Musa textilis* produced a lower kappa number compared to *Eucalyptus pellita*.

The delignification degree and the amount of kappa pulp are significantly affected by active alkali. Figure. 1 shows the effect of the amount of active alkali on the level of delignification and the amount of kappa from the pulp produced. The delignification degree increases with the increasing amount of active alkali. As a result, the amount of kappa pulp decreases with increasing amount of active alkali. At an active alkaline amount of 20%, the delignification rate reaches around 92% for *Eucalyptus pellita* and 88% for abaca. The concentration of $OH^-$ and $HS^-$ in liquor significantly influences the level of delignification. It is well understood that HS- increases the cell wall swelling and facilitates delignification. The relatively low kappa rate of 13-16 (at 20% active alkali) is achieved with an H-factor of 800. It seems that Abaca is better prepared than *Eucalyptus pellita* wood because abaca kraft pulp requires an H-800 factor to reach a kappa pulp number of 10,15.

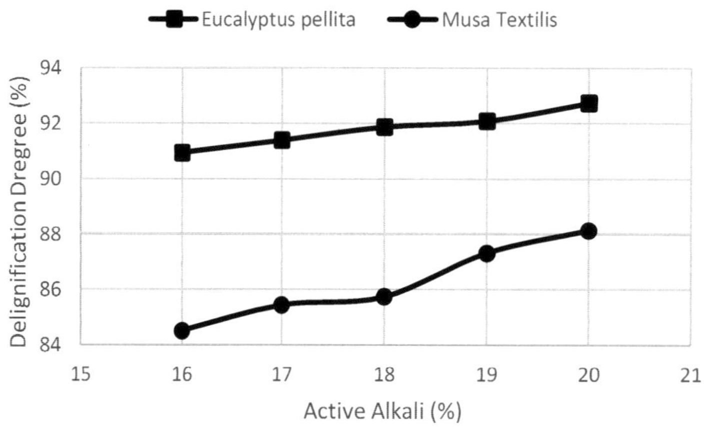


**Figure 1.** Effect of active alkali amount on delignification degree

The higher the viscosity, the higher the carbohydrate resistance. The nature of good physical endurance depends on the bond between fibers [10]. Table. 4 shows that viscosity is low with increasing active alkali load. High viscosity levels on active alkali loads are low because of the density or bonding between strong fibers, even though cellulose is also degraded by liquids. A high active alkali load will produce low viscosity due to high polymerization rates. Bonding between fibers, especially cellulose, is degraded by liquor, resulting in low fiber strength and high pulp solubility [4]. The high viscosity of pulp is expected to improve pulp quality, especially pulp and tensile strength. Viscosity and kappa numbers are used to determine the content and loss of degradation of lignin and carbohydrates in the pulping process. Lignin is widely lost and cellulose degradation increases with increasing active alkali content used in the pulping process. Lignin is lost by breaking the bond between the α- and β-aryl ether bonds in the propane unit. Subsequent termination by the liquid at the hydrolysis temperature decreases the amount of cellulose.

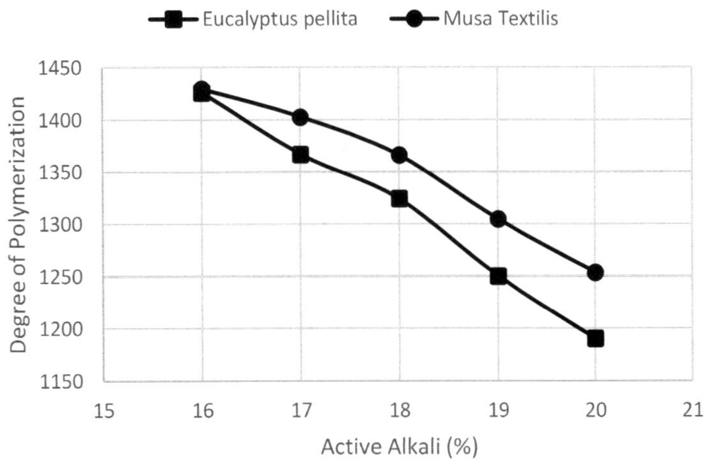


**Figure 2.** Effect of Active alkali on Degree of polymerization

As with cellulose, the decrease in hemicellulose causes a decrease in the viscosity of the pulp caused by a reduced degree of polymerization [11]. Loss of hemicellulose at a low degree of delignification can produce a higher viscosity than the pulp. A decrease in viscosity of a paper pulp that cannot be removed by more than 20% can increase the content of $OH^-$ which decreases cellulose lignin. The degradation of cellulose by alkali can be continued by peeling off the reaction and hydrolysis of the alkali [12].

*3.2 Oxygen delignification*

The kappa number, yield, and viscosity of the kraft pulps are presented in Table 5 before and after oxygen delignification. It can be seen that the degree of delignification at the same sodium hydroxide load depends on the initial kappa number [11]. For pulp with an initial kappa number 10-16, the delignification rate is between 30 and 38% and a final kappa number between 7 for pulp kraft abaca and 12 for pulp kraft eucalyptus is obtained. These results indicate that kraft pulp with low kappa number can also be easily removed by oxygen delignification [12][5].

The lower the kappa number after cooking, the lower the kappa number after oxygen reduction. The results showed that the lignin content in the oxygen-delignification stage pulp one level could be reduced to a low level either by reducing lignin residues during cooking or by increasing the alkali content at the oxygen delignification stage. However, pulp with lower lignin content (kappa number 7-9) and great viscosity can only be obtained if the amount of lignin residue is quite low after cooking.

Great viscosity of pulp after oxygen removal has a degree of polymerization for pulp of ≥ 1200 [13] (according to ISO 5351: 2012 standards) except for pulps cooked with AA 19-20% shown in figure 3. At the final kappa number 9.2-10, 62 in *Eucalyptus pellita*, the highest viscosity in the pulp case of AA 16% is 861.98 ml/g together with a low alkaline load (2%) in the oxygen delignification stage. At the final kappa number of 7.04-8.4, the highest viscosity for abaca kraft pulp at AA 16% is 886.02 ml/g higher than that of the *Eucalyptus pellita* kraft pulp. In the last kappa number around 7, the viscosity is quite low. It can be seen from this result that delignification kraft pulp to kappa numbers as low as 7 by one-stage oxygen delignification

**Table 5.** Properties of the kraft pulps before and after oxygen delignification

| Material | Active Alkali (%) | Initial Kappa Number | Kappa Number Final | Initial Viscosity (ml/g) | Viscosity Final (ml/g) | Yield (%) |
|---|---|---|---|---|---|---|
| *Eucalyptus pellita* | 16 | 16.69 | 10.62 | 954 | 861.98 | 96.46 |
| | 17 | 15.87 | 10.36 | 918 | 838.45 | 95.67 |
| | 18 | 14.99 | 9.68 | 892 | 820.32 | 95.73 |
| | 19 | 14.6 | 9.37 | 847 | 778.42 | 95.77 |
| | 20 | 13.4 | 9.2 | 810 | 748.55 | 94.72 |
| *Musa textilis* | 16 | 13.24 | 8.4 | 955.82 | 886.02 | 95.01 |
| | 17 | 12.45 | 7.62 | 939.68 | 870.50 | 95.73 |
| | 18 | 12.2 | 7.48 | 917.50 | 833.43 | 95.56 |
| | 19 | 10.85 | 7.20 | 880.17 | 795.94 | 94.80 |
| | 20 | 10.15 | 7.04 | 848.98 | 768.83 | 95.66 |

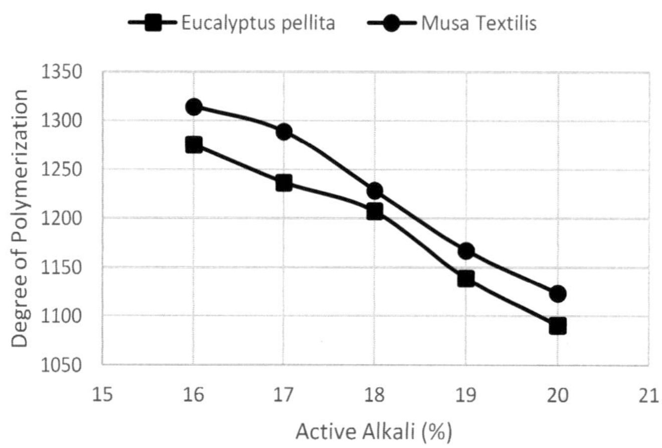


**Figure 3.** Effect of Oxygen delignification on Degree of polymerization

results in a severe reduction in pulp strength, as demonstrated by the low viscosity observed [3]. It has thus been proven to be possible to abolish selected pulp kraft abaca by oxygen delignification to pulp with very low kappa numbers (7-9) and great viscosity (degree of polymerization $\geq$ 1200). Based on the data, abaca fiber is more effective for delignification of lignin polymers than *Eucalyptus pellita.* This result is very interesting for modern pulp mills, which aim, for example, to use ECF or TCF bleaching techniques. However, more work needs to be done to clarify the bleaching capability, and especially the feasibility of the pulp strength, of this pulp with low residual lignin content.

## 4. Conclusions

In this study, abaca as delignified by kraft pulping and oxygen-alkali delignification subsequently becomes a pulp with a low kappa amount. The most important findings from various different alignment experiments are a higher alkaline load can increase alkaline residues and reduce yield, kappa number and viscosity and the active alkali (AA) has a significant effect on the kappa number and viscosity of kraft pulp obtained. Lower AA (6-18%) results in higher viscosity and pulp yields compared to higher

AA (19-20%) at the level of delignification. Kraft pulp with high cooking results and good viscosity is easily delignified by converting oxygen to low kappa numbers (7-9) without significant loss in viscosity. A higher level of viscosity obtained from low AA compared to high AA was also observed in oxygen-depleted pulp. In addition, from the final results, the kappa number and viscosity, it was shown that abaca pulp has very high strength. Although modified pulp and oxygen delignification for lower kappa quantities can be considered, in terms of the properties of the pulp the abaca fiber is suitable as a raw material for pulp and has the potential to substitute *Eucalyptus pellita.*

## Acknowledgment

This work has been financially supported by a research grant of "Penelitian Dasar Unggulan Perguruan Tinggi (PDUPT)" for year of 2019 from DRPM, Indonesia, to which the authors express their gratitude.

## References

[1] Santoso B, Mastur and Kadarwati F T 2016 Abaca (Musa textilis Nee) As The source of Natural fiber, Producing raw Material for Pulp and Source of Farmer Income *Perspektif* **15** 1–10
[2] Kumar Sinha A, Narang H K and Bhattacharya S 2017 Effect of Alkali Treatment on Surface Morphology of Abaca Fibre *Mater. Today Proc.* **4** 8993–6
[3] Li J, Zhang C, Hu H and Chai X-S 2016 Effect of shortening kraft pulping integrated with extended oxygen delignification on biorefinery process performance of eucalyptus. *Bioresour. Technol.* **202** 119–24
[4] Júnior E A B and Gomes F 2018 The effects of temperature , alkali charge and additives in the oxygen delignification in high kappa number eucalyptus pulp kraft *Sci. For.* **118** 217–27
[5] Zhao H, Li J and Zhang X 2018 Fundamental understanding of distracted oxygen delignification efficiency by dissolved lignin during biorefinery process of eucalyptus *Bioresour. Technol.* **258** 1–4
[6] Rutkowska E W, Wollboldt P, Zuckerstätter G, Weber H K and Sixta H 2009 Characterization Of Structural Changes In Lignin During Continuous Batch Kraft Cooking Of Eucalyptus Globulus *BioResources* **4** 172–93
[7] Vijayalakshmi K and Kavitha A 2014 Abaca Fibre *Trans. Eng. Sci.* **2** 16–9
[8] Edson T, Segura S and Sarto C 2016 Effect of Kappa Number Variation on Modified Pulping of Eucalyptus *BioResources* **11** 9842–55
[9] Bajpai P 2018 *Nonwood Fiber Use in Pulp and Paper* (Elsevier)
[10] Gabriela A, Carvalho M, Duarte F A S, Gomes F and Sacon M 2001 The effect of alkali charge on Eucalyptus spp . kraft pulping
[11] Rahmati H, Ebrahimi P and Sedghi M 2010 Effect of cooking conditions and oxygen-delignification on Bambusa tulda kraft pulping *Indian J. Chem. Technol.* **17** 74–7
[12] Liu Z, Cao Y, Yao H and Wu S 2013 Oxygen delignification of wheat straw soda pulp with anthraquinone addition *BioResources* **8** 1306–19
[13] Przybysz Buzała K, Przybysz P, Kalinowska H and Derkowska M 2016 Effect of Cellulases and Xylanases on Refining Process and Kraft Pulp Properties. *PLoS One* **11** e0161575

IOP Conf. Series: Materials Science and Engineering **857** (2020) 012022 doi:10.1088/1757-899X/857/1/012022

# Enhanced cloud based mobile core network with network function virtualization

**H D Septama[1,*], A Ulvan[1], M Pratama[1]**

[1] Department of Electrical Engineering, University of Lampung, Jalan Prof. Soemantri Brojonegoro No.1, Bandar Lampung 35145, Indonesia

*Email: hery@eng.unila.ac.id

**Abstract.** The mobile technology is improved by adopt the Internet Protocol (IP) based technology. The IP Multimedia Subsystem or IP Multimedia Core Network Subsystem (IMS) is a technology designed by 3rd Generation Partnership Project (3GPP). This technology used IP to deliver the communication packet and recognized as Next Generation Network (NGN). The emerging IP based technology took a lot of advantages by using IP to deliver a packet. However, some issues i.e. availability, security and reliability still need consideration to meet the telecommunication standard. The cloud based computing become buzzword as a technology for cost reduction and hardware efficiencies. This paper conducted the feasibility to setup mobile core network in cloud based environment and enhanced it with network function virtualization (NFV) feature. The NFV made the network more scalable, efficient and robust network management. The QoS is presented as the results to concluded the feasibilities. The results show that cloud based mobile core network with NFV is feasible since the QoS still full fill the telecommunication standard.

**Keywords:** core network, NFV, cloud, mobile

## 1. Introduction

The communication nowadays become human basic need. The mobile communication subscribers increase each year with various demand of communication. The voice communication then become multimedia communication. User use not only voice and messaging but also use video and internet access. The mobile technology also improved by adopt the Internet Protocol (IP) based technology. The IP Multimedia Subsystem or IP Multimedia Core Network Subsystem (IMS) [1] is a technology designed by 3rd Generation Partnership Project (3GPP). This technology used IP to deliver the communication packet and recognized as Next Generation Network (NGN). The signalling is delivered with Session Initiation Protocol (SIP) [2] and for the media is using Real Time Protocol (RTP) [3]. The emerging IP based technology took a lot of advantages by using IP to deliver a packet. However, some issues i.e. availability, security and reliability still need consideration to meet the telecommunication standard.

The deployments of telecommunication technology always using dedicated devices. Since the emerging telecommunication technology also using an IP based, therefore we may deploy the service in the cloud environments. The cloud based computing become a use full technology for cost reduction and hardware efficiencies. The telecommunication companies may reduce capital expenditures

(CAPEX) and operating expenditures (OPEX) since a lot of part is running virtually. The cloud computing is believed as one of future telecommunications infrastructures [4]. However, since IP is not a reliable communication protocol, the deployment should ensure that the QoS of telecommunication standard is fulfilled.

This paper conducted the feasibility to setup mobile core network in cloud based environment enhanced with network function virtualization (NFV) feature. The QoS is presented as the results to concluded the feasibilities.

## 2. Related works

The works in [5] virtualized network elements based on LTE SAE/EPC core network. The paper deployed network elements in virtualized environments and make some QoS measurements to ensure the feasibilities. The testbed consisted of several virtual machine running networks element service and make a voice call to the VoIP server. Traffic generator also used to simulate the actual traffic condition. The results show the networks element works well in virtualization environment.

The works in [6] explain why we need to setup the IMS mobile core networks over cloud networks. The works show that using cloud computing technology will help to make IMS more scalable and easy to deploy. The paper also proposed the possibility of the IMS using Infrastructure as a Service (IaaS) and Platform as a Service (PaaS) and what their weakness.

The NFV made revolutionary changes on network implementation by made the network features running virtually. The leading telecommunication companies support and implement the NFV and is believe become one of the main key of future network technologies [7]. The use cases and standard landscape of NFV is standardized, the work in [8,9] show several cases of how the NFV could help driving a transition towards future programmable carrier grade networks. Thus, replace the default bridge interface with NFV not only made the infrastructure more scalable and efficient but also easier to managed.

## 3. Cloud and NFV architecture

Cloud computing is an emerging computing technology. This technology enables computing scalability dynamically by virtualized its hardware resources. Users may access cloud through various devices i.e laptop, computer, smart phones. The advantages using cloud technology is cost savings, hardware efficiencies, high availability and easy scalability [10].

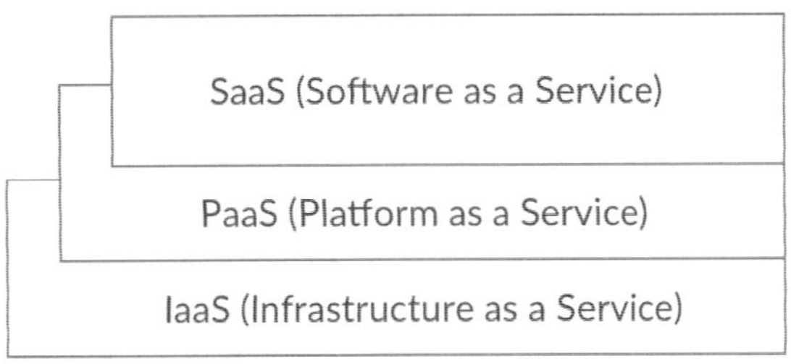


**Figure 1.** Cloud computing layer

Figure 1 depicts the layer of cloud computing architecture by the services its offered. The top layer is Software as a Service (SaaS), this kind of services is offered an application to be running remotely by the users. Second layer is Platform as a Services (PaaS) that offered a ready to use platform for he users. It means providers should provide not only the hardware resources but also an operating system equipped with software stack to be able users installed or run its own applications. The last layer is Infrastructure as a Services (IaaS), the services that only provide the infrastructure or hardware resources and user manages the infrastructure itself. There are three types of cloud, public cloud that provide its service access through global internet. Private cloud that deployed its cloud on a private networks and combination of both types that called hybrid cloud.

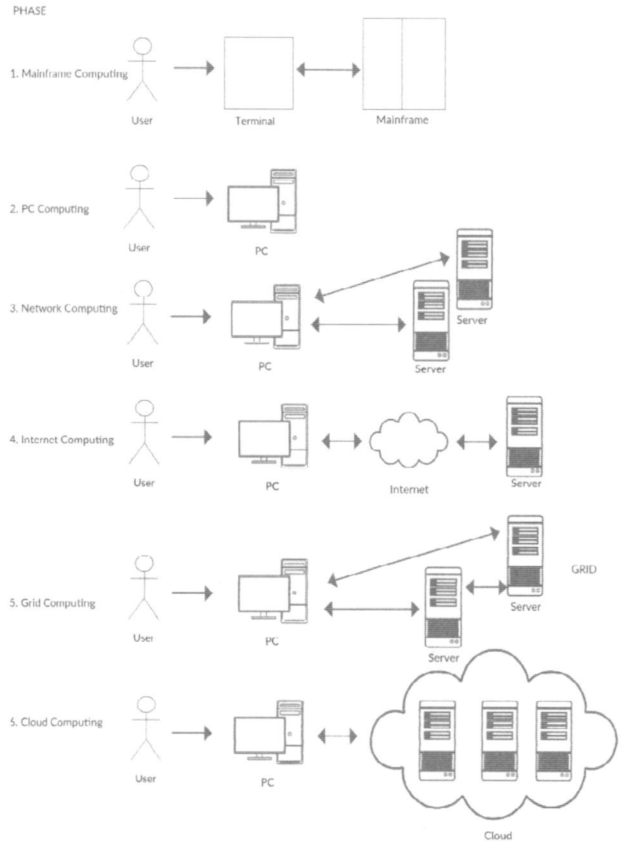


**Figure 2.** Six computing paradigms [4]

Figure 2 depicts the six phase of computing paradigms. Each phase shows the paradigm or method to access and how allocate the computer resources. The cloud computing method become novel computing architecture. This architecture become widely used since its offers various advantages such as cost reduction and scalable environments. Users only pays that they need or used for the time being, as for the demand is increased the amount of computer resources may be added to the services. However, the cloud also has introducing another problem such are security and latency. A lot of research is done to provide more secure and less latency for the cloud network service.

The ETSI high level NFV framework is depicted on figure 3. The figure shows the relation of hardware resources, virtualization layers/hypervisor and virtualised network function that runs on top of it. NFV management and orchestration is used to manage and control the virtual network functions therefore the management and control process become centralized and easier. The addition, deletion, configuration and another interfaces management and control features of virtual network function is done by single application.

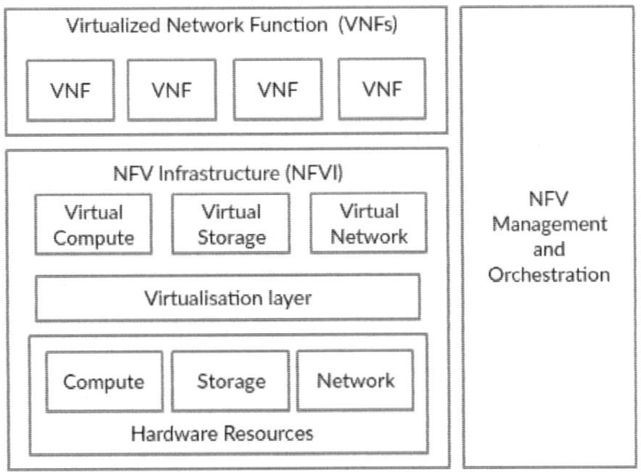


**Figure 3.** High-level NFV framework [4]

## 4. Testbed and scenarios

This paper use private cloud infrastructure for the testbed deployments. Figure 4 is the testbed high level architecture model. The mobile core network elements are deployed as a virtual application over operating system in a virtual machine (VM).

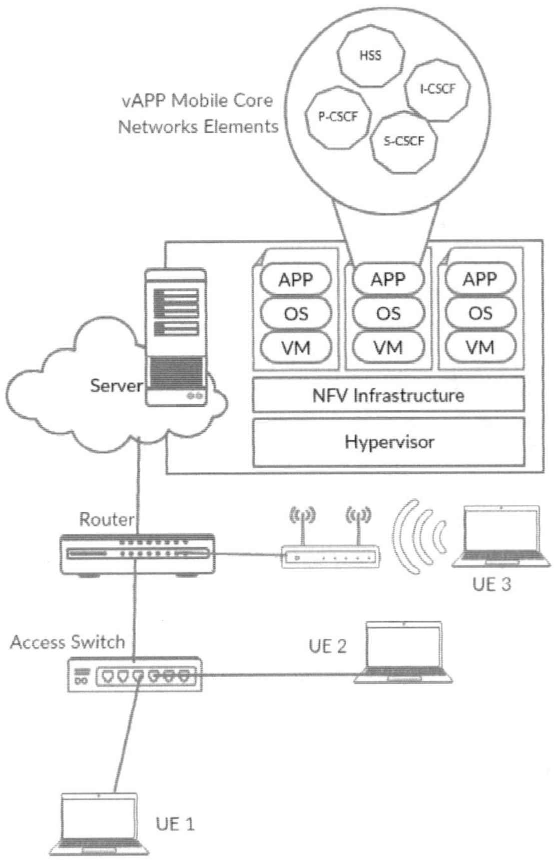


**Figure 4.** The tesbed environments

IOP Conf. Series: Materials Science and Engineering **857** (2020) 012022 doi:10.1088/1757-899X/857/1/012022

The VM is connected with virtual interface using NFV instead of using its default bridge configuration. The main reason to use the NFVI is to provide fully manage and control to user using centralized managements plane. Thus, two user equipment (UE) is connected via wired connection and another one UE is connected via wireless (Wi-Fi) connection. UE 1 is also equipped with packet sniffing software to capture the packet during call scenarios procedure. The sniffed packet is use to analysis the QoS of the communication to conclude the feasibility using mobile core network in cloud based environments.

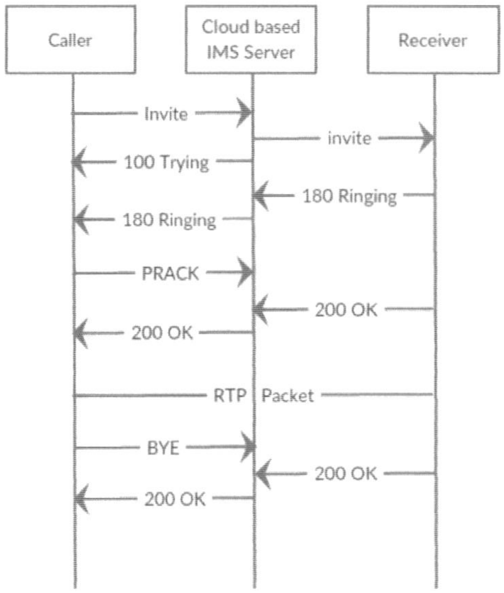


**Figure 5.** Call scenarios procedure

There are several scenarios conducted to observed the QoS feasibility. Firstly, the voice call quality for wired access connection are observed between UE 1 and UE 2. Secondly, the voice call for wireless communication quality also observed between UE 1 and UE 3 communications. Last, the video call communication quality between UE 1 and UE 2. All call is using SIP signalling protocols and G.711 codec for voice communication. Figure 5 show the basic call procedure for the call test. During the call, RTP packet is sniffed and then the Jitter calculated to present the quality. RTP is a protocol defined by IETF RFC 3550 [3] that provide real time data for delivery service.

| 2 | 3 | 4 | 8 | 9 | 16 | 32 |
|---|---|---|---|---|---|---|
| V | P | X | CSRC Count | M | Payload type | Sequence number |
| Timestamp | | | | | | |
| Synchronization source (SRC) | | | | | | |
| Contributing source (CSRC: variable 0 - 15 items, 2 octets each | | | | | | |

**Figure 6.** RTP packet structure

Figure 6 depicts the RTP packet structure, the timestamp field is used to identify each packet for jitter calculation purposes. This field contains the sampling instant of the first octet in the RTP data

IOP Conf. Series: Materials Science and Engineering **857** (2020) 012022 doi:10.1088/1757-899X/857/1/012022

packet that derived from a clock and incremented monotonically and linearly in time to allow synchronization and jitter calculations. Jitter of this RTP packet is calculated using jitter calculation described in IETF RFC 3550 [3].

## 5. Result and Discussion

In this paper, inter arrival jitter calculation is considered to analysed the performance. However, the delay and packet loss for each scenario also presented. The ITU – T standard in table 1 [10] describes parameters for good telephony communication i.e., the average MOS of 4.4, delay < 150 ms, jitter < 20 ms and packet loss < 5 %.

**Table-1**. ITU-T standard for delay and jitter

| Grade | Delay (ms) | Jitter (ms) |
|---|---|---|
| Very good | 0 | 0 |
| Good | 0 - 150 | 0 – 20 |
| Average | 150 - 30 | 20 - 5 |
| Poor | >300 | >50 |

Inter arrival jitter (J) and mean deviation of the difference (D) defined for pairs packet, by comparing packets spacing at the receiver and sender. Packet spacing is the difference between the RTP packet timestamp from sender and receiver time when arrived packet also knows as relative transit time.

$$D(i,j) = (Rj - Ri) - (Sj - Si) = (Rj - Sj) - (Ri - Si) \tag{1}$$

Where:
*Si,Sj* = RTP timestamp from packet i and j,
*Ri,Rj* = Time of arrival in RTP timestamp units for packet i and j

Equation 1 above showed the difference (D) calculation of two packets i and j. Since the RTP packet arrived continuously at the receiver then Difference (D) from pairs of packets (n packet and n – 1 packet) also calculated continuously. Therefore, the inter arrival jitter could be calculated using equation 2 as follows:

$$J(i) = J(i-1) + \frac{(|D(i-1,i)| - J(i-1))}{16} \tag{2}$$

The Inter arrival jitter algorithm above is the optimal first-order estimator and 1/16 gain parameter used in order to give a good noise reduction ratio while maintaining a reasonable rate of convergence.

The RTP packet sniffed during the call scenarios then calculated. The calculated results is depicted in figure 7 through 8 below. Figure 7 show the inter arrival jitter of voice call for first scenarios procedure using wired access connection. The results show that max jitter is 14,42 ms with mean jitter 13,91 ms. The packet loss is 0% with delta max or latency 40,5 ms. Figure 8 show the jitter of voice call for wireless connection. The results show that that max jitter is 14,26 ms with mean jitter 13,47 ms. The packet loss is 0% with delta max or latency 41,67 ms. As for the video call depicted in Figure 9, the max jitter is 15,86 ms with mean jitter 13,99 ms. The packet loss is 0% with delta max or latency 50,76 ms.

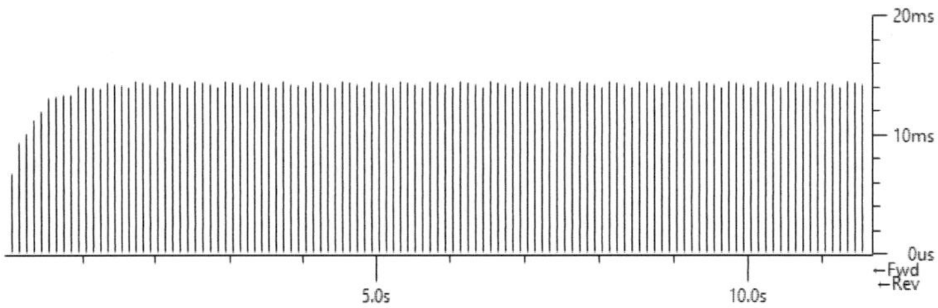


**Figure 7.** Jitter wired voice call

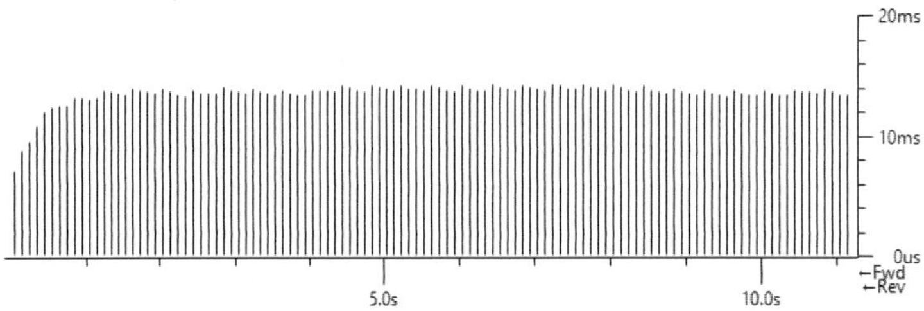


**Figure 8.** Jitter wireless voice call

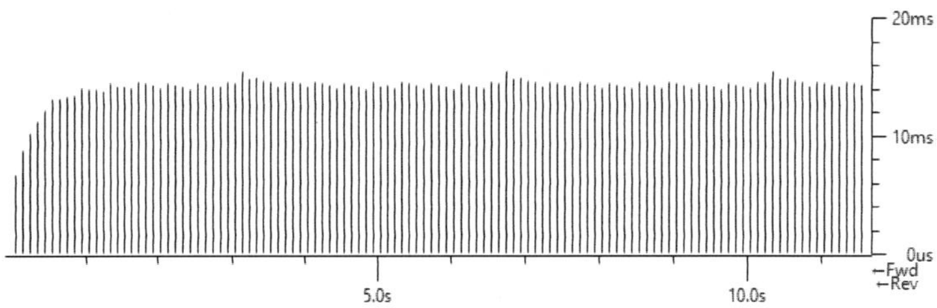


**Figure 9.** Jitter video call

## 6. Conclusions

The results show that the QoS of cloud based mobile core network with NFV is feasible. The jitter, packet loss and latency is still fit the ITU.T recommendation in a good grade. For future works the call generator considered to deployed for providing traffic load, so that the simulation results can represent the actual state of the network when operated.

## References

[1] 3GPP TS 33.401, "The 3rd Generation Partnership Project; Technical Specification Group Services and System Aspects; 3GPP System Architecture Evolution (SAE): Security Architecture (Release 8)", 3GPP, 2009.

[2] Rosenberg, J. et. al. SIP: Session Initiation Protocol. IETF RFC 3261. Internet Engineering Task Force (IETF). 2002

[3] Schulzrinne, H. Casner, S. Frederick, R. Jacobson, V. RTP: A Transport Protocol for Real-Time Applications. IETF RFC 3550. Internet Engineering Task Force (IETF). 2003

[4] Furht, B. Escalante, A. et all. 2010 Handbook of Cloud Computing (New York: Springer)

[5] Ulvan A, Melvi, Bestak R and Septama, HD 2017 J. Engineering and Applied Sciences ARPN **12** 365-370

[6] Glitho, R. 2014 Proc. 6th Int. Conf. on New Technologies, Mobility and Security (NTMS) (Dubai: IEEE) p 1
[7] Hawilo H, Shami A, Mirahmadi M and Asal R 2014 J. Networks IEEE **28** 18
[8] Kozicki B, Olaziregi N, Oberle K, Sharpe R B and Clougherty M 2014 Proc. Globecom Workshops (Austin: IEEE) p 595
[9] ETSI Group Specification NFV, Network Function Virtualization (NFV) use cases, ETSI GS NFV 001, European Telecommunications Standards Institute (ETSI), 2013
[10] One-way transmission time. ITU.T Recommendation G.114. International Telecommunication Union Telecommunication Standardization Sector (ITU.T). 2003

# Effect of sintering holding time on the properties of Hydroxyapatite granules fabrication using dripping technique of HA-Alginate sintered at 1000 °C

**M D Effendi[1,*], D Gustiono[1]**

[1] Agency for The Assessment and Application of Technolgy, Gedung 224, Jl. Kw. Puspiptek, Muncul, Kec. Setu, Kota Tangerang Selatan, Banten 15314, Indonesia

[*] Email: dachyar.effendy@bppt.go.id

**Abstract**. CaP granules as bone graft are used for open defect of bone, where is not necessary mechanical stability they are proposed for no-load bearing applications. The important feature in the physical structure of a synthetic ceramic bone graft is its porosity, pore size, the volume of porosity and interconnections between the pores are three crucial parameters. The pore structure is of great importance for osteoconduction. The level of porosity, pore size distribution and degree of pore interconnectivity significantly influence the extent of bone ingrowth. According to that, this study investigated the effect of sintering holding time on properties of droplet synthetic ha-alginate using dripping technique sintered at 1000 °C with sintering holding time at 1: 2 :3 hours respectively. XRD was used to investigate of phases of the sintered droplet. Scanning Electron Microscope (SEM) was used to investigate the microstructure of beads. In this work, we can conclude that the holding time increased, the crystallinity increased but on the other hand, pore size, porosity, and surface roughness decreased. For all investigated samples. The surface roughness of A102 (0.2 – 2, µm) which is sintered at 1000 °C 2 hours holding time has a better value offered better cell colonialization and has no significant value of porosity (50%) and crystallinity (86.63%) than those of A103 which need more energy for sintering as a consequence.

**Keywords :** Hydroxyapatite, CaP, sintering, microstrucure, surface roughness, granule

## 1. Introduction

CaP has been known as the material for bone grafting caused by its similarity component with the bone [1], [2]. One of Cap, hydroxyapatite, is accepted as bioactive for guided bone regeneration. On the other hand, hydroxyapatite has a weakness in poor mechanical properties such as low compressive strength and fracture toughness [3]. So that, hydroxyapatite has considered applying in non-load bearing part for implant material [4,5] and applied in the form of powder, granule and porous scaffold where is the porous scaffolds are highly attractive due to good osteoconductivity and fast bone ingrowth [6]. Compared to porous scaffolds, the granule hydroxyapatite is easier to produce.

The scaffold has a clearer shape and big macroporosity (typical range 50-500 µm) as a big role in osteoconductivity mechanism compared to granules such as osteoconductivity and fast bone ingrowth, due to high surface area and sufficient blood circulation such as osteoconductivity and fast bone ingrowth, due to high surface area and sufficient blood circulation. However, scaffolds need to cut and press-fitting into the defect. That is the problem of using scaffolds. So that, for smaller open defects, using granules is more attractive than using scaffolds. Some researcher stated that Microstructure of the bone graft ceramic material has main role of the recovery of bone defect , such as surface roughness is influence to colonization of cell [7] and correlation between surface roughness and osteoconduction has been well described in several systematic reviews [8–10].So that, the roughness surface was also evaluated in this work.

Granules hydroxyapatite can be produced with some ways : planetary mixers [11], fluidized-bed granulators [12,13], spheronizers [14,15], pelletizers in forming granules from wet masses containing the desired ingredients [15–17]and recently, droplet extrusion or beads formation [18,19]. This method is based on a gelification mechanism involving natural polysaccharides. Hydrogels are three-dimensional hydrophilic tissue that can absorb large amounts of water or biological fluid, mimics biological tissue [20]. Polymer gel beads are made using a number of natural biodegradable polymers. Because they come from natural sources, do not require organic (toxic) solvents, are easily available and of high quality for a number of chemical modifications [21–23] . Polyelectrolyte polymers are used in the manufacture of hydrogel matrix beads, the most widely used is alginate [24,25] and chitosan [26–29], and also carboxymethyl cellulose [30–32] have been applied.

Ribeiro at al [25] was mixed with sodium alginate solution (3% w/v) at a ratio of 0.2 w/w and well homogenized, they were dried overnight in a vacuum-oven at 30°C, and then sintered at 1100°C for 1h, with a uniform heating rate of 5°C/min from room temperature. This previous research did not declare about the influence of holding time of sintering temperature at 1000 $^{0}$C on its properties. In this work, we investigated the influence of holding time of sintering at 1000 $^{0}$C on its properties of HA granules produced by droplet extrusion technique and determine the properties of sintered granules such as value of crystallinity, pore size, phase identification, and surface roughness as the biological requirement for bone graft material.

## 2. Materials and Methods

### *2.1 Materials*

1) *Materials preparation*

   Hydroxyapatite powder obtained from process as follow: 1M solution of calcium oxide and 0.6 gr/lt [$(NH_4)_2HPO_4$] is used as the initial solution. The solution of $(NH_4)_2HPO_4$ added drop-wise in a solution of calcium oxide. Ammonia was used for adjustment of pH values. The suspension is stirred with a magnetic stirrer after the addition of the initial solution for 24 hours at room temperature. Precursor done for 30 minutes at room temperature, then filtered and washed with a vacuum Buchner for five times until a neutral pH is reached. For the first and third phases were washed with distilled water and ammonia, the second phase and the fourth was washed with distilled water to remove residual dirt, and last washed with ethanol to remove water and increase the dispersibility. The end product was dried at room temperature for 24 hours then heated to 160$^{0}$C for 17 hours for the removal of water. Sintering process was carried out at 650$^{0}$C with a heating rate of 80 minutes with a time of 2 hours' detention. hydroxyapatite samples then characterized by XRD.

2) *Droplet formation.*

   Droplet extrusion method allows the production of spherical drops starting from a bioceramic suspension containing Alginate solution that allows the instantaneous jellification in a crosslinking solution. HA powder has added into composition of microspheres CaP-alginate gelation was mixed with sodium alginate solution (Sigma Aldrich 3% w/v) at a ratio of 0.2 w/w and well homogenized. The paste was extruded drop-wise into a 0.1M $CaCl_2$ cross-linking solution, where spherical-shaped particles instantaneously formed and were allowed to harden for 30 min. The size of the microspheres was controlled by regulating the extrusion flow rate using a syringe. At completion of the gelling period, microspheres were recovered and rinsed in water in order to remove the excess $CaCl_2$. Finally, they were dried overnight in a vacuum-oven at 30ºC, and then sintered at 1100ºC for 1h, 2h and 3 h with a uniform heating rate of 5ºC/min from room temperature.

*2.2 Characterization techniques*

The sintered Hap granules were investigated with following characterization techniques. To characterize the crystallized solids chemical analyses, powder X-ray diffraction (XRD), FT-IR spectroscopy, thermal analysis and SEM methods were applied. Philips X'pert XRD equipment with a graphite monochromator (PW 1752/00), CuKα, (λ = 0.15418 nm) radiation, a Ni filter in 2× range of 10 to 60 degrees at 30 kV, 30 mA was used.

Thermal decomposition was investigated using a TGA-DTA (simultaneous Thermogravimetric Analysis (TGA) and Differential Thermal Analysis (DTA) fully automated system (Universal V2.3C TA Instruments) with 20 deg/min heating rate. The SEM images were obtained using JEOL JSM 7500F. The total and apparent density and total porosity of sintered compacts, shrinkage and weight loss during the sintering were also measured. Total density was determined by simple pycnometric method. Hydrostatic weighing (HW) uses the Archimedes principle was used to calculate apparent density of sintered specimens. The procedure of HW involves using the following three measurable values: the weight of the body outside the water, the weight of the completely immersed body and the density of the water. The value of total and apparent density is needed to evaluate the percentage of total material porosity (P) according to the equation:

$$P = 1 - \frac{appereance\ density}{total\ density}\ x\ 100\% \quad (1)$$

## 3. Results and discussion

*3.1. XRD analysis of synthetic HA*

Figure 1 shows the result of precipitation method to synthesize hydroxyapatite as describe as above. We used Highscore plus® to identify the pattern of powder sample.

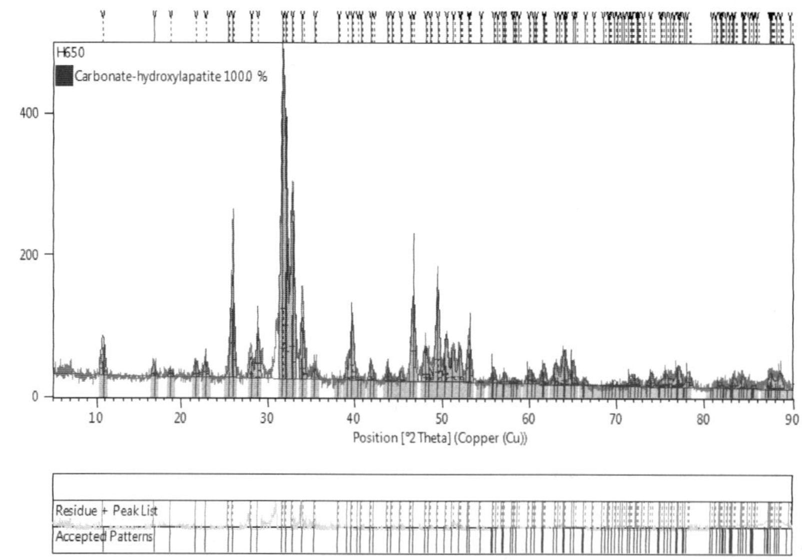


**Figure 1.** The result of search and Match using high score plus software with Rietveld analysis of the synthetic HA powder sintered at 650 $^0$C

Diffraction pattern was match to Carbonate-hydroxyapatite / $Ca_{10.00}P_{6.00}O_{26.14}H_{2.60}C_{0.02}$ (Card. No. 96-900-3554). Table 1 shown the lattice parameter has no significant different with the lattice parameter of the reference. It can conclude that the sample phase was Carbonate-hydroxyapatite. On the sintering temperature at 650$^0$C, the HA powder has 539-696Å and 64% degrees of crystallinity.

*3.2. XRD analysis of sintered granules*

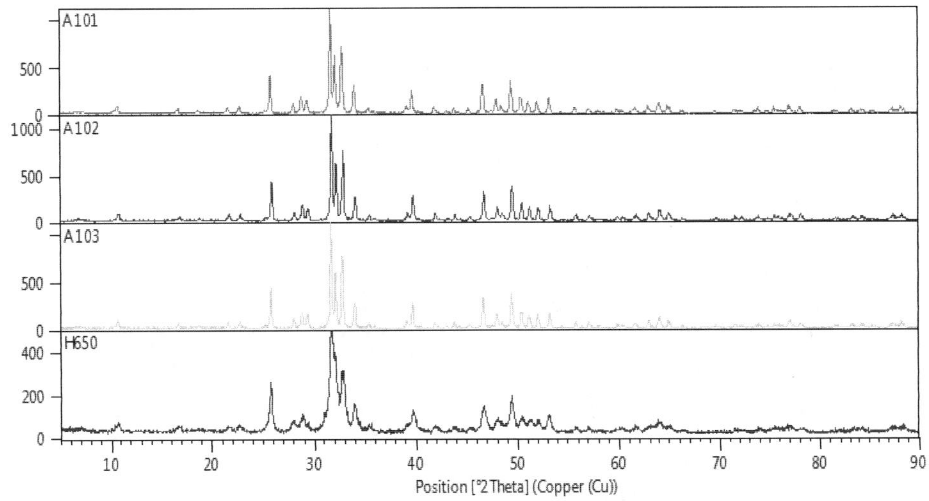


**Figure 2.** XRD pattern of all samples sintering at 1000°C with holding time at 1, 2, 3, hours respectively.

From figure 2. Shown that all of sintered samples has diffraction pattern identically. No phase transformation found compare to HA powder and between all samples. All of pattern had similar to diffraction pattern of Carbonate-hydroxyapatite / $Ca_{10.00}P_{6.00}O_{26.14}H_{2.60}C_{0.02}$ (Card. No. 96-900-3554) which has sharper peak at 32$^0$.

**Table 1**. Crystallography properties

| Sample Code | c | a | Crystallite size , Å | % crystallinity | Intensity at highest peak | FWHM |
|---|---|---|---|---|---|---|
| HA powder, 650 $^0$C | 9.4109 | 6.8777 | 137-212 | 64 | 203 | 0.5088 |
| Reff. Carbonate-hydroxyapatite Card #96-900-3554 ($Ca_{10.00}P_{6.00}O_{26.14}H_{2.60}C_{0.02}$) | 9.4470 | 6.8810 | - | | | |
| A101 | 9.4468 | 6.8806 | 359-762 | 86.34 | 565.06 | 0.1886 |
| A102 | 9.4316 | 6.8760 | 361-745 | 86.63 | 591.77 | 0.1893 |
| A103 | 9.4279 | 6.8742 | 393-869 | 86.08 | 600.15 | 0.1738 |

The crystallography of sintered granules and sintered powder was evaluated and the result are shown at table 1. No phase transformation detected due to the lattice parameter for each sample. The crystallinity increased when the holding time of sintering increased, but slightly decreased for 3 hours holding time (A103). The linear relationship is shown between the holding time and crystallite size where the holding time increased, the crystallite size increased.

*3.3. SEM analysis*
Results for SEM data were presented on Fig. 3 to Fig. 5 below

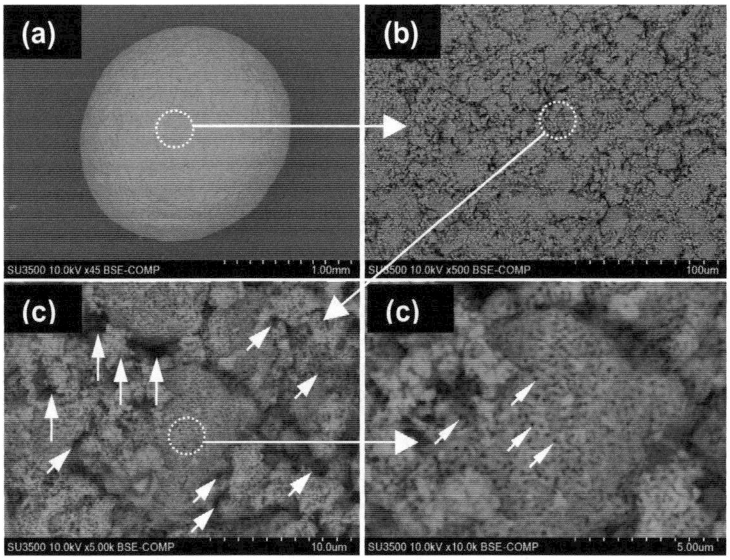


**Figure 3.** SEM of sintered granules (A101). Sintering Temperature at 1000 $^{0}$C, holding time 2 hour with magnifications 45x, 500x, 5k, and 10k respectively

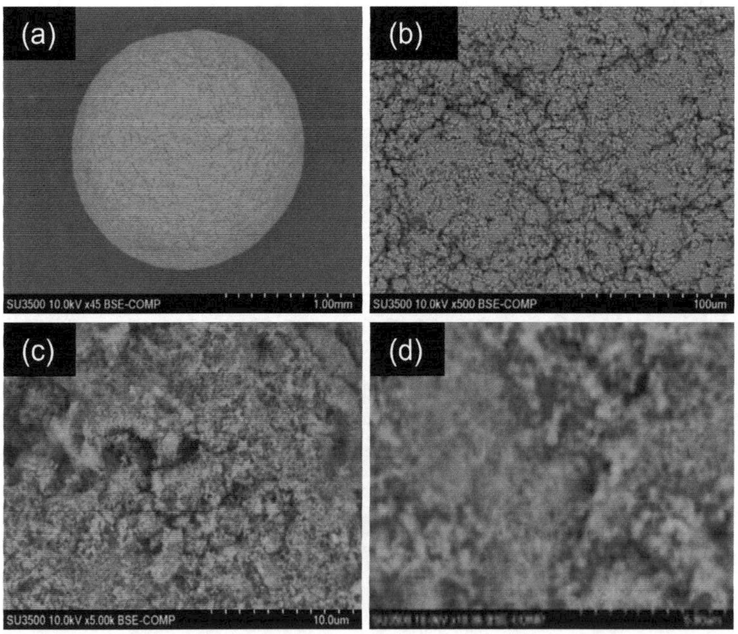


**Figure 4.** SEM of sintered granule (A102). Sintering Temperature at 1000 $^{0}$C, holding time 2 hour with magnifications 45x, 500x, 5k, and 10k respectively

**Figure 5.** SEM of sintered granule (A103). Sintering Temperature at 1000 $^0$C, holding time 2 hour with magnifications 45x, 500x, 5k, and 10k respectively

**Table 2**. Properties of the granules

| Sample Code | Surface roughness, μm | Pore size, μm | Porosity, % |
|---|---|---|---|
| A101 | 0.25-2.1 | 0.12 - 0.4 | 51 |
| A102 | 0.2-2 | 0.13-0.4 | 50 |
| A103 | 1-28 | 0.2 – 3.2 | 50 |

Based on the Table 2 it can declared that the holding time on sintering process had influenced on pore size, degrees of porosity and surface roughness where surface roughness and porosity decreased relatively when the holding time increased caused by densification. On the other hand, pore size decreased when holding time increased. The surface roughness become smoother and less of porosity. It is in agreement with Patel, 2001 [33]. At the contrast, in A103 we found a large pore size and we considered it could be effect of non-homogenous mixing proses between alginate and HA slurry for investigated sample.

## 4. Conclusions

Hydroxyapatite sintering holding time has a significant impact on surface roughness, on microstructure including crystallinity and micro porosity were shown to be of great importance in promoting protein adsorption on the material surface and capillary effect, and would consequently influence cell adhesion, proliferation, and bone tissue growth. So that, the fulfillment of properties of microstructure of bone graft for that biological requirement is depending of sintering process such as degree of temperature and holding time. This work shown that at the same degree of sintering temperature, the holding time effected on surface roughness, crystallinity, pore size, and porosity. It is known that smooth surfaces are less expected to conduct cell colonization; for that reasons, the holding time for 2 hours at 1000 °C is better than other and not significant different of pore size and degrees of porosity of A101. Besides, the crystallinity is better than holding time at 1 hour. The crystallinity of a biomaterial is important caused by high crystallinity materials display lower biodegradation rates and better volume stability. It means that dissolution of material is depend of degree of crystallinity. In 2 hours holding time at 1000 $^0$C sintering temperature, we obtained 86.63% crystallinity compare to 3 hours holding time (86.08%) but with the less energy for sintering process.

**Acknowledgment**
This work was supported by Project of Inovation of biocompatible materials for implant, BPPT 2019.

**References**

[1] S. Shahgoli and M. H. Levine, "Introduction and overview of bone grafting.," *N. Y. State Dent. J.*, 2011.
[2] N. Eliaz and N. Metoki, "Calcium phosphate bioceramics: A review of their history, structure, properties, coating technologies and biomedical applications," *Materials (Basel).*, 2017.
[3] H. A. Siddiqui, K. L. Pickering, and M. R. Mucalo, "A review on the use of hydroxyapatite-carbonaceous structure composites in bone replacement materials for strengthening purposes," *Materials*. 2018.
[4] M. R. Mucalo, *Hydroxyapatite (HAp) for Biomedical Applications*. 2015.
[5] "The Chemistry of Medical and Dental Materials," *Oral Dis.*, 2003.
[6] P. Chocholata, V. Kulda, and V. Babuska, "Fabrication of scaffolds for bone-tissue regeneration," *Materials*. 2019.
[7] D. D. Deligianni, N. D. Katsala, P. G. Koutsoukos, and Y. F. Missirlis, "Effect of surface roughness of hydroxyapatite on human bone marrow cell adhesion, proliferation, differentiation and detachment strength," *Biomaterials*, 2000.
[8] A. Wennerberg and T. Albrektsson, "Effects of titanium surface topography on bone integration: A systematic review," *Clinical Oral Implants Research*. 2009.
[9] H. H. H. Hessam, S. I. S. Izman, and S. H. S. Hamtaiepour, "Evaluating the surface properties of hydroxyapatite coating on titanium alloy subtrate," *J. Mek.*, vol. 36, no. 1, 2013.
[10] M. M. Shalabi, A. Gortemaker, M. A. Van't Hof, J. A. Jansen, and N. H. J. Creugers, "Implant surface roughness and bone healing: A systematic review," *Journal of Dental Research*. 2006.
[11] A. Faure, I. M. Grimsey, R. C. Rowe, P. York, and M. J. Cliff, "A methodology for the optimization of wet granulation in a model planetary mixer," *Pharm. Dev. Technol.*, 1998.
[12] M. Dosta, S. Heinrich, and J. Werther, "Fluidized bed spray granulation: Analysis of the system behaviour by means of dynamic flowsheet simulation," *Powder Technol.*, 2010.
[13] P. Roy, R. Khanna, and D. Subbarao, "Granulation time in fluidized bed granulators," *Powder Technol.*, 2010.
[14] M. Thommes and P. Kleinebudde, "The Science and Practice of Extrusion-Spheronization," 2017.
[15] M. Evers, D. Weis, S. Antonyuk, and M. Thommes, "Scale-up of the rounding process in pelletization by extrusion-spheronization," *Pharm. Dev. Technol.*, 2019.
[16] P. Nold, R. Löbe, and M. Müller, "Granule production - Easy and cost-effective," *InterCeram Int. Ceram. Rev.*, 2003.
[17] S. Shanmugam, "Granulation techniques and technologies: Recent progresses," *BioImpacts*, 2015.
[18] B. V. Parakhonskiy *et al.*, "Size controlled hydroxyapatite and calcium carbonate particles:

Synthesis and their application as templates for SERS platform," *Colloids Surfaces B Biointerfaces*, 2014.

[19] M. Descamps, J. C. Hornez, and A. Leriche, "Manufacture of hydroxyapatite beads for medical applications," *J. Eur. Ceram. Soc.*, 2009.

[20] Q. Chai, Y. Jiao, and X. Yu, "Hydrogels for Biomedical Applications: Their Characteristics and the Mechanisms behind Them," *Gels*, 2017.

[21] M. R. Singh, S. Patel, and D. Singh, "Natural polymer-based hydrogels as scaffolds for tissue engineering," in *Nanobiomaterials in Soft Tissue Engineering: Applications of Nanobiomaterials*, 2016.

[22] L. Gasperini, J. F. Mano, and R. L. Reis, "Natural polymers for the microencapsulation of cells," *Journal of the Royal Society Interface*. 2014.

[23] M. C. Catoira, L. Fusaro, D. Di Francesco, M. Ramella, and F. Boccafoschi, "Overview of natural hydrogels for regenerative medicine applications," *J. Mater. Sci. Mater. Med.*, 2019.

[24] J. Sun and H. Tan, "Alginate-based biomaterials for regenerative medicine applications," *Materials*. 2013.

[25] C. C. Ribeiro, C. C. Barrias, and M. A. Barbosa, "Calcium phosphate-alginate microspheres as enzyme delivery matrices," *Biomaterials*, 2004.

[26] M. D. Effendi, D. Gustiono, D. Ayu, and F. Kurniawati, "Comparison on mechanical properties of single layered and bilayered chitosan-gelatin coated porous hydroxyapatite scaffold prepared through freeze drying method," *IOP Conf. Ser. Mater. Sci. Eng.*, vol. 172, p. 12031, Feb. 2017.

[27] N. Bhattarai, J. Gunn, and M. Zhang, "Chitosan-based hydrogels for controlled, localized drug delivery," *Advanced Drug Delivery Reviews*. 2010.

[28] M. C. G. Pellá, M. K. Lima-Tenório, E. T. Tenório-Neto, M. R. Guilherme, E. C. Muniz, and A. F. Rubira, "Chitosan-based hydrogels: From preparation to biomedical applications," *Carbohydrate Polymers*. 2018.

[29] A. Anitha *et al.*, "Chitin and chitosan in selected biomedical applications," *Progress in Polymer Science*. 2014.

[30] N. A. Zakharov, Z. A. Ezhova, E. M. Koval, V. T. Kalinnikov, and A. E. Chalykh, "Hydroxyapatite-carboxymethyl cellulose nanocomposite biomaterial," *Inorg. Mater.*, 2005.

[31] D. Pasqui, P. Torricelli, M. De Cagna, M. Fini, and R. Barbucci, "Carboxymethyl cellulose - Hydroxyapatite hybrid hydrogel as a composite material for bone tissue engineering applications," *J. Biomed. Mater. Res. - Part A*, 2014.

[32] M. Sayed, H. F. El-Maghraby, F. Bondioli, and S. M. Naga, "3D carboxymethyl cellulose/hydroxyapatite (CMC/HA) scaffold composites based on recycled eggshell," *J. Appl. Pharm. Sci.*, 2018.

[33] N. Patel, I. R. Gibson, S. Ke, S. M. Best, and W. Bonfield, "Calcining influence on the powder properties of hydroxyapatite," *J. Mater. Sci. Mater. Med.*, 2001.

# Implications of automotive product sustainability on young customers' purchase intention in developing countries: an experimental approach

**A Y T Panuju[1], D A S Ambarwati[2] and M D Susila[1]**

[1] Dept. of Mechanical Engineering, Faculty of Engineering, University of Lampung, Indonesia
[2] Dept. of Management, Faculty of Economy and Business, University of Lampung, Indonesia

* E-mail: yahyateguh@gmail.com

**Abstract.** Manufacturing of sustainable products is on a high demand in order to attain a global responsible consumption and production pattern. Among the emerging efforts of various manufacturing companies, production of sustainable vehicles takes a special attention due to its massive impacts to the world's sustainability dimensions. Meanwhile, youngsters in developing countries are interesting prospective customers for automotive products because their misterious characteristics and potential to contribute to the future sustainability. The research investigated the connectivities between automotive products sustainable designs and young customers purchase intention in Indonesia, using an experimental workshop. Empirical results have shown that each dimension has a significant effect to the youngsters' preferences, although with different levels.

**Keywords:** purchase intention, sustainable products, automotive, customer behaviour

## 1. Introduction

Since the responsible consumption and production pattern has become one of the pillars for a global sustainability, sustainable product development has been a concern for manufacturing industries in recent years [1,2]. Therefore, paradigm in product development area has shifted to a direction with more environmental and social considerations in developing new products, which subsequently are mentioned as sustainable products. Several methods to design these sustainable products have been proposed and have led to the creation of new products in many aspects of human life [3-5]. Among these aspects, transportation vehicles one of the most common example where sustainability concept has been widely implemented due to its significant impact to the environment and society [6-9].

As the sustainable vehicles are proliferating in markets around the world, responses obtained in different countries may vary. The most significant gap can be seen in developed countries and developing countries, due to disparities found in the customer behaviour [10]. These disparities may come from different level of education and prosperity, so that the awareness level of sustainability also differs. Regarding the size of markets in developing countries is unneglectable [11], it is crucial to develop more understanding about correlations between customers' behavior and sustainable automotive products in developing countries to develop further strategies directing to the global sustainability.

Furthermore, we agree that strategic policies of related industries should be made based on observations through young people in age 18 – 30 [12-13]. As one of the reasons, young customers buying intention toward sustainable products reflects the opportunity of sustainable living style in the future [14-15]. We have seen how these young people give great influence in consumption pattern during these years, and this milennial behavior is still moving extremely fast, sometimes to unpredictable directions. Further, regarding the milennials as potential market for automotive products in the future, it is essential for the designers to pay attention on what attracting them to purchase sustainable cars or motorcycles. By taking the tripple bottom line (TBL) as the dimensions of sustainability [16], it is interesting to find whether the environmental concern, the social impact, or economic reasons affect the young customers most in taking the purchase decision. Therefore this study focuses on discussion about how young customers in a developing country (Indonesia as a case study) respond to products' sustainability dimensions (TBL) in a manner of purchase intention. The results can provide useful resources for the designers and other interested scholars to understand young customer behavior in designing sustainable products for developing countries with similar characteristics.

## 2. Literature Review

### *2.1. Automotive Product Sustainability*

Sustainability in one product is assessed through all its life cycle, considering several category / dimension of sustainability. These dimensions are subsequently divided into criteria reflecting more specific condition (table 1). Since this research discusses sustainability in customer's perspective, criteria discussed in this research are limited to those only with direct benefits to the customer.

**Table 1.** Criteria in Automotive Product Sustainability Assesment (Customer's View)

| Category | Criteria | Criteria's code | Refferences |
|---|---|---|---|
| Environmental Impact | Global warming and ozone depletion | E1 | [6,17,18] |
| | Toxicity and pollution to the environment | E2 | [6,17,18] |
| | Scarcity of resources | E3 | [6,17,18] |
| Social Impact | User's Health | S1 | [6,17,18] |
| | User's Safety | S2 | [6,17,18] |
| | User's Comfort (vibration, noise) | S3 | [6,17] |
| | User's Mobility (performance and durability) | S4 | [6,17,18] |
| Economic Impact | Operating Cost | M1 | [6,17,18] |
| | Maintenance Cost | M2 | [6,17,18] |
| | Used product value | M3 | [17,18] |

Refferences about vehicle sustainability assessment methods are not many, however several studies have been accomplished to develop the methods. Each methods present different criteria, although having similarities. Ungureanu [6] proposed a framework consisting six category, constituting triple bottom line with functionality, manufacturability and recyclability as additional points. Functionality had many similar functions with social criteria, while manufacturability and recyclability sub criteria can be categorized in environment criteria. Shuaib et al [18] still used TBL in his method, which can be used for general products, to score the product's sustainability index. Salvado [7] also used TBL in his work, generating sustainability index using analytic hierarchy process. Meanwhile, Jasinski [17] argue resource impact as another criteria inspite of TBL, because automotive products are recognized as one of the most resource-consuming products used by the society. However, it is assumed from the customers' point of view, resource consumption also means environmental sake. For practical reasons, criteria used in this discussion uses basic TBL in product operating phase.

### *2.2. Product Sustainability and Consumers'Purchase Intention*

Purchase intention of customers is a measuring value for predicting future sales of a certain product, although it is not the only factor to be considered in marketing [19]. Researches related to the field are enormous with different aims and approaches , and those which relates to product sustainability are sufficiently provided as well . Understanding customer purchase intention is a crucial part of planning future sustainability, because achieving sustainable consumption pattern also partially means selling sustainable products to as many customers as possible around the world.

The problem is, although the sustainable products sales trend recently shows an inclination [20-21], its percentage among other conventional products trend is still low. Customer's reluctancy mostly comes from unawareness, lack of information and poor habits toward sustainability (Tang 2008). However, by increasing customer's perceived value of the product and knowledge about related risks, customer's purchase intention is aimed to be higher [22]. While knowledge is wished to be obtained by education and advertisements, products' perceived value should be build integrated in the product design [23]. Incorporating sustainability dimensions in a product requires a comprehensive design method, which is limited by several boundaries including technology availability, customer behaviour considerations, ruling policies and the effects of the design to sustainability dimensions. Though the process is surely more complex and more cost demanding than the old conventional product design methods, applying these sustainable methods is a necessity to attain excellences in nowadays competitive markets [24].

Correlations between product sustainability dimensions and customer's purchase intention has been investigated in prior studies, and some of them are given here. Chen [25] investigated whether an eco-motorcycle design affect consumers' purchase intentions in Taiwan. The research has proven that customer's knowledge holds significant role in increasing purchase intention. Toppinen [26] found that consumer's preferences to sustainable wood products in Finnish should be divided into dimensions: environment and social. Park [27] has proposed a concept using the TBL framework to conceive customer's perception in fashion industry.

Many studies have been conducted in developed countries, while consumers in developing countries are recognized having different characteristics than those who live in developed countries [28]. A few samples of studies conducted in developing countries related to the discussion are mentioned here. Handriana [29] examine three groups constituting professionals, housewives and young people in Indonesia as a developing country. All groups showed positive response to green products, even each has different level of familiarity. Interesting participants in this research were the young people, which represented potential of future sustainability in developing countries. They seem to have more interest in global issues. For instance, in comparison to housewives, young people are more familiar and having higher buying intention to green products. However, Shamsi [30] found that the customer's educational backgrounds is the most significant aspect in determining the response to green products. It is affected by the amount of information gained by the prospective customers, and how this information interact with the customer's knowledge related to sustainability.

## 3. Methods

### *3.1. Hypotheses Development*

Ghazali et al [31] investigated cultural influences on green product preferences in Indonesia. Among the cultural traits presented, authors found the salient factors influencing the market are uncertainty avoidance and long-term orientation. Further, it can be concluded from the discussion that the society has pragmatical point of view about green products, referring to money savings oriented. This is supported by Lim [32] who revealed that economic incentive has significant influence in Malaysian antecedents' purchase intention, while the environmental issue does not. Chen [33] found that perceived monetary value could highly seduce consumers to buy green products, but environmental awareness was also considered by consumers in purchase decision making. However, this pragmatical financially related mind is not seen in young consumers. Lian and Yoong [34] have revealed that young consumers in Malaysia prefer to consider social dimension (health and safety) in comparison to affordability. From another study, Kowang [35] stated functionality of vehicles influents youth's purchase intention significantly in Malaysia, reflecting the youngsters have high preferences to the engine performance and the vehicles durability. It shows that youngsters relatively have attention to social dimension sustainability like health, safety and products functionality. This behavior is easily understood as consumers' basic consideration in buying products, also applied for older community. For instance, in certain types of product, health concern enggaged with sustainability issue, got the highest attention from prospective customers in an online market website [36]. While Hung [37] found that vehicle safety awareness among consumers is very high more than economic reasons, which can influent their buying intention to the related products. These findings might lead to a hypotheses that social dimension would be the top reason for young customers to buy green vehicles. While economic dimension might be the less consideration, since the respondents in the proposed workshop would be assumed financially adequate to buy the variant models.

Meanwhile, environmental impact consideration is still unfamiliar for several market segments, the youngsters have recognized the issue quite well [24]. This is because young customers are widely opened to new information and affected easily by media [38]. For these milennials, getting a comprehensive understanding about environmental issues is not a difficult thing to do since there are many advertisings related to environmental issues in media recently. In prior researches, it has been clarified that a better level of knowledge is considered to increase the purchase intention on eco products among young customers [30]. In addition, Setyawan [39] found that even though currently for most Indonesian adolescents environmental issues have not been a main topic in their mind, their purchase intention to green products might increase significantly by applying adequate information delivery. These findings are similar with Zhang's research [14] in China with green houses as the purchasing objects. He acclaimed that young customers in China are willing to buy green houses as long as they get sufficient explanation about urgency of the green products. According to these findings, the following hypotheses are proposed.

**H1**: Social dimension of sustainability has a significant impact to young customers purchase intention and it has higher impact in comparison to other dimensions.
**H2**: Environmental dimension of sustainability has a significant impact to young customers' purchase intention which is lower than the social dimension, but higher than economic dimension.
**H3**: Economic dimension of sustainability has significant impacts to young customer's purchase intention, but has the lowest influence compared to other dimensions.

### *3.2. Workshop as an Experimental Tool*

Relation between sustainable features in automotive product and customers' purchase intention should be examined by eliminating influences from other factors outside the variables such as brand loyalty, pricing, etc. For this research, it is decided to enroll experimental research, in form of a workshop. Experimental method have been used widely in measuring consumer's behavior to sustainable products in various way. Some of them measured the customer's willingness to pay [40-41]. The method is acclaimed to decrease the bias possibility due to the preliminary treatments to the participants because it consists explanation session from the researchers. In this case, there are two main subjects to be described to the respondents at the beginning of the workshop, which are a brief explanation of sustainability concept and the characteristics of the product models used in the research.

As the experiment object, motorcycle variants were chosen as the product models to be presented in the workshop due to their wide usage among youngsters in Indonesia. Three imaginary motorcycles with different characteristics were developed to be explained to the respondents in a certain sequence (table 2), then the respondents were asked to fill questionnaires related to the products (Q1) and their intention to buy them (Q2). There are 10 questions in each questionnaire constituting the sustainability criteria as listed in table 1. Likert scale 1 to 5 was used to analyze whether the respondents are strongly agree, agree, normal, disagree or strongly disagree with the statements.

Each motorcyle variant represents excellences in environmental, social and economic dimension respectively, so that the respondents' purchase intention to each variant should depict the respondents' preferences to sustainability dimensions. Other factors which might effect the respondents decision, such as the bike shape, brand, price, colors, are set to be flexible according to the respondents' taste.

**Table 2.** Motorcycle Variants as Product Models and Their Characteristics

| | Criteria's Code | Performance Defined | | Criteria's Code | Performance Defined | | Criteria's Code | Performance Defined |
|---|---|---|---|---|---|---|---|---|
| **Variant 1** | E1 | Excellent | **Variant 2** | E1 | Poor | **Variant 3** | E1 | Poor |
| | E2 | Excellent | | E2 | Poor | | E2 | Poor |
| | E3 | Excellent | | E3 | Poor | | E3 | Poor |
| | S1 | Poor | | S1 | Excellent | | S1 | Poor |
| | S2 | Poor | | S2 | Excellent | | S2 | Poor |
| | S3 | Poor | | S3 | Excellent | | S3 | Poor |
| | S4 | Poor | | S4 | Excellent | | S4 | Poor |
| | M1 | Poor | | M1 | Poor | | M1 | Excellent |
| | M2 | Poor | | M2 | Poor | | M2 | Excellent |
| | M3 | Poor | | M3 | Poor | | M3 | Excellent |

During the workshop, respondents should be grouped into classes, where each class should contain at least 30 respondents for the sake of measurement validity. These classes are required to avoid the testing factor where a certain sequence in explaining things in a workshop might have effects to the respondents' responses. In avoiding such a phenomena, in each class the products explanation sequences would be set differently. Figure 1 is presented as a model to understand how the workshop would be conducted.

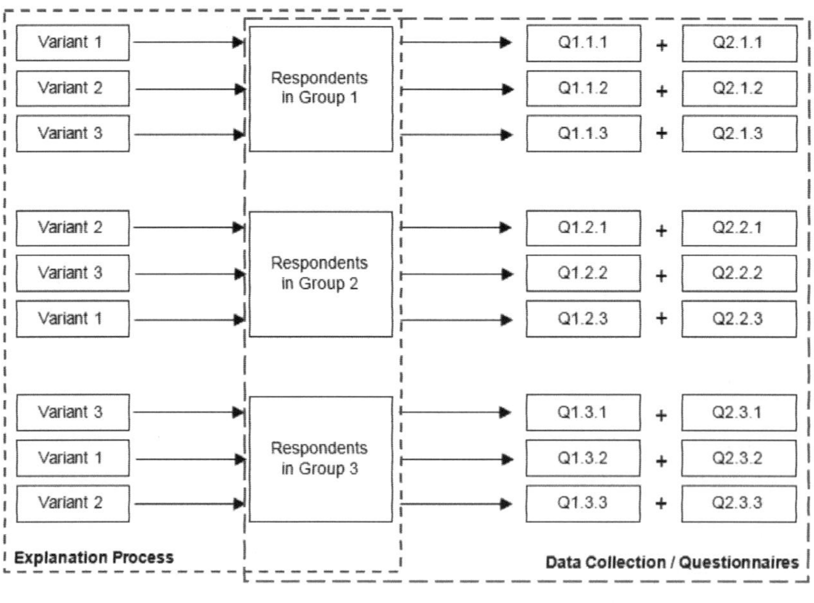


**Figure 1.** The Experimental Workshop Model

Workshops in these classes were conducted simultaneously with the same materials and questionnaires, only with different respondents and sequence of product explanation. Since Q1 and Q2 questionnaire each contains 10 questions, thus totally there are 303 respondents' responses for the three variant models. These data subsequently are analyzed by using statistical tools to reveal the respondents purchase intention to each variants.

## 4. Results and Discussions

The experimental workshop has been conducted in August 2019, where 101 respondents were involved. The profiles of the respondents are given in table 3.

**Table 3.** Respondents Profiles

| Respondents Profiles | | Freq | % |
|---|---|---|---|
| Gender | Male | 59 | 58,4 |
| | Female | 42 | 41,6 |
| Education Level | High School | 42 | 41,6 |
| | Bachelor | 45 | 44,5 |
| | Master | 14 | 13,9 |
| Age | 18 – 21 | 51 | 50,4 |
| | 22 – 26 | 40 | 39,6 |
| | 27 – 30 | 10 | 10 |

An experimental work requires less respondents in comparison to a full survey work. It has relatively stronger internal validity, though it suffers weaker external validity [35]. KMO and Bartlett's tests have been implemented to the data and have confirmed that the measuring instrument has fulfilled the threshold level (KMO > 0.5; all factor loadings > 0.5; Bartlett's should be significant with $p < 0.05$). In addition, by analyzing communalities and factor loadings, it is described that each dimension of TBL has specific significant effect to the customer purchase intention. Therefore for each dimension' the cronbach's alpha has been counted and validated above the required level (Cronbach's alpha > 0.7). The results are given in table 4.

**Table 4.** Validity and Reliability Tests Results

| Measurements | | |
|---|---|---|
| Kaiser-Meyer-Olkin sampling adequacy tes | | 0.793 |
| Bartlett's Test of Sphericity | Approx Chi Square | 3129.217 |
| | Df | 45 |
| | Sig | 0.000 |
| Cronbach's alpha (environmental dimension) | | 0.919 |
| Cronbach's alpha (social dimension) | | 0.970 |
| Cronbach's alpha (economic dimension) | | 0.896 |

Once the validity and the reliability of the instruments has been validated, it is essential to ensure that the respondents' perception to the presented products is correct. Questionnaire Q1 has been provided for this intention, then a manipulation check using T-test, means value and significancy has been enrolled to reveal the perception. As the result, most of respondents clearly understand the characteristics and the excellences of each variant. For examples, the respondents confirmed that variant 1 is the best for its environmental dimension characteristics while the variant 2 has the best performance in social dimension, and so on. All data have significancy below 0.05, so it is considered that the respondents' perception is sufficient. It is assumed that the experimental workshop method is an appropriate approach to reveal young customers behavior due to its ability to clarify the problems to the audience at the beginning of the workshop. Thus the measurement of their purchase intention should be more valid under this condition, as on Table 5.

**Table 5.** Homogenity Test

| Classifications | Age | Education Level | Gender |
|---|---|---|---|
| Asymptotic Sig. (Pearson Chi Square) | 1.000 | 1.000 | 1.000 |

Subsequently, the respondents' profiles should be examined whether these factors affect the purchase intention or not. There are three classifications in the profiles which should be noticed: age, education level and gender. Using crosstabulations and chi square tests for the 3 classifications versus the 3 variants, data in table 5 has been collected. It shows that none of these classification has a noticable effect to the respondents' purchase intention, because it requires value under 0.05 for the asymptotic significance to be considered as significant. While other studies proposed that difference in education background should affect customers preferences, it is assumed that a brief good explanation as conducted in the workshop could eliminate this knowledge limitation among the respondents [33, 40-41].

**Table 6.** Univariate Analysis of Variance

| | (I) code | (J) code | Mean Difference (I-J) | Std. Error | Sig. | Subset 1 | Subset 2 | Subset 3 |
|---|---|---|---|---|---|---|---|---|
| Tukey HSD & Bonferroni | Variant 1 | Variant 3 | 2.9307 | 0.85454 | 0.002 | 29,10 | | |
| | | Variant 2 | -2.8218 | 0.85454 | 0.003 | | | |
| | Variant 2 | Variant 1 | 2.8218 | 0.85454 | 0.003 | | 31,93 | |
| | | Variant 3 | 5.7525 | 0.85454 | 0.000 | | | |
| | Variant 3 | Variant 1 | -2.9307 | 0.85454 | 0.002 | | | 26,17 |
| | | Variant 2 | -5.7525 | 0.85454 | 0.000 | | | |

**Model Summary**

| Model | R | R Square | Adjusted R Sq | Std. Error of the Estimate |
|---|---|---|---|---|
| 1 | 0.765 | 0.616 | 0.615 | 4.03002 |

The connectivities between the product sustainability dimensions to the young customers' purchase intention has been investigated on data collected through questionnaire Q2, using univariate analysis of variance. By using between-subject tests and multiple comparisons based on Tukey HSD and Bonferroni methods, the results clarify that all sustainability dimensions have specific effects to the customers' purchase intention, although with different level. It is shown by the significance value, with all value are below 0.05. This result should be a good sign that sustainable motorcycles might have better responses from Indonesian customers in the future.

Furthermore, the analysis from homogenous subset reveals that the social dimension has the highest implication on the purchase intention, followed by the environmental and economic dimension respectively. The results fulfill the hypotheses, where the young customers do not focus on financial benefits from the products, but they prefer to satisfy their personal needs first and contribute to the environmental issues instead. Overall, the sustainability dimensions in the automotive product models have affected the young customers' purchase intention by 61% compared to other reasons. This result validates the hypotheses described in previous section (table 6).

## 5. Conclusion

An investigation to young customers' purchase intention to sustainable vehicles has been conducted using an exprimental workshop in Indonesia. Three motorcycle models with specific characteristics have been used to test the preferences of 101 respondents. As the result, it can be seen that young customers in Indonesia have main interests on social dimension criteria in purchasing a motorcycle, constituting health, safety, comfort and product functionality considerations. Oppositely, these young customers put economic dimension, including used product value, operational and maintenance cost as the lowest consideration. One can argue that the condition is caused by the assumption made at the beginning of the workshop, where the respondents were assumpted financially adequate. Thus the respondents' preference may differ in the real life condition, when facing economical boundaries. On the other side, environmental dimension considerations may have a good opportunity to be highly considered by customers in the future, regarding good education and advertisements about the urgency of taking care of the nature. Even in developing countries, environmental issue shall be a good added value for products to attract more customers, because the awareness of sustainability among youngsters have been existed.

## Acknowledgments

This research is funded by DIPA Faculty of Engineering University of Lampung, Indonesia. Authors thank all related parties contributing to the research, especially to Zaky Abyan Farhain as the authors' research assistant in conducting the workshop.

## References

[1] Myklebust O 2006 Sustainable Industrial System *Intelligent Strategies in Product Design, Manufacturing and Management* ***(207)****: 637 – 642*

[2] Clark G, Kosoris J, Hong L N, Crul M 2009 Design for Sustainability: Current Trends in Sustainable Products Design and Development *Journal of Sustainability* ***1****: 409 – 424*

[3] Vinodh S, Rathod G, 2010 Integration of ECQFD and LCA for Sustainable Product Design, *Journal of Cleaner Production* **18**:833 – 842

[4] Hassan M F, 2011 Methodology for Sustainable Product Design: A Review and Direction of Research *International Conference on Mechanical and Manufacturing Engineering* Malaysia

[5] Kobayashi H, 2015 Perspectives on Sustainable Product Design Methodology Focused on Local Communities *Proceedings of EcoDesign International Symposium*

[6] Ungureanu C A, Jawahir I S, Das S 2007 Development of a Sustainability Scoring Method for Manufactured Automotive Products: A Case Study of Auto Body Panels *Proceedings of IMECE USA*

[7] Salvado M F, Azevedo S G, Matias J C O, Ferreira L M 2015 Proposal of a Sustainability Index for the Automotive Industry *Journal of Sustainability* ***(7)****: 2113 – 2144*

[8] Drohomeretski E, da Costa S E G, de Lima E P, Neves T R O 2015 The Application of Sustainable Practices and Performance Measures in the Automotive Industry: a Systematic Literature Review *Engineering Management Journal* ***(27)1****: 32 – 44*

[9] Vaz C R, Rauen T R S, Lezana A G R 2017 Sustainability and Innovation in the Automotive Sector: a Structured Content Analysis *Journal of Sustainability* ***9***
[10] Tamura T, Kobayashi H, Umeda Y 2016 Requirements for Design Methodology for Local Oriented Manufacturing in Developing Countries *Procedia CIRP* ***40****: 431-436*
[11] OECD 2012 Green Growth and Developing Countries – A Summary for Policy Makers
[12] Naderi I, Steenburg E V 2018 Me First, and then The Environment: Young Millennials as Green Consumers *Business Faculty Publications Fairfield University*
[13] Palupi T, Sawitri D R 2018 The Importance of Pro-Environmental Behavior in Adolescent *EDP Sciences*
[14] Zhang L, Chen L, Wu Z, Zhang S, Song H 2018 Investigating Young Consumers' Purchasing Intention of Green Housing in China *Journal of Sustainability* ***10***
[15] Zahid W, Dastane O 2016 Factors Affecting Purchase Intention of South East Asian Young Adults towards Global Smartphone Brands *Asean Marketing Journal (8)****1****: 66 – 84*
[16] Elkington J, 1999 Cannibal with Forks: Triple Bottom Line of 21st Century Business *Capstone*
[17] Jasinski D, Kirwan K, Meredith J 2016 A Comprehensive Framework for Automotive Sustainability Assessment *Journal of Cleaner Production* ***(135)****: 1034 – 1044*
[18] Shuaib M, Seevers D, Zhang X, Badurdeen F, Rouch K E, Jawahir I S 2014 Product Sustainability Index (ProdSI) *Journal of Industrial Ecology* ***(18)4****: 491 – 507*
[19] Morwitz V 2012 Consumers' Purchase Intentions and their Behavior *Foundations and Trends in Marketing* ***(7) 3****: 181-230*
[20] Srivastava M, Malik K, Kumar R 2016 Consumers' Readiness for Green Products and Its Effect on Sustainable Development *Amity Journal of Marketing. ADMAA* **1(1):** 93 – 104
[21] Kim H, Lee C W 2018 The Effect of Customer Perception and Participation in Sustainable Supply Chain Management: A Smart Phone Industry Study *Journal of Sustainability. MDPI* **20**
[22] Chen H C, Lin M H 2018 A Study on The Correlation Among Product Design, Statistics Education, and Purchase Intention – A Case of Toy Industry *EURASIA Journal of Math. Science and Technology Education.* **14(4):** 1189 – 1195
[23] Yen H Y, Lin P H, Lin R 2014 Emotional Product Design and Perceived Brand Emotion *International Journal of Advances in Psychology (IJAP)* ***3(2)****: 59-66*
[24] Seuring S 2010 Supply Chain Management for Sustainable Products *Business Strategy and The Environment: 471 – 484*
[25] Chen H S, Tsai B K, Hsieh C M, 2017 Determinants of Consumers' Purchasing Intentions for the Hydrogen-Electric Motorcycle. *J. Sustainability* ***9***
[26] Toppinen A, Toivonen R, Valkeapaa A, Ramo A K 2015 Consumers Perception of Environmental and Social Sustainability of Wood Products in the Finnish Market *Scandinavian Journal of Forest Research* ***28(8)****: 775-783.*
[27] Park H, Kim Y K 2016 An Empirical Test of The Triple Bottom Line of Customer-Centric Sustainability: The Case of Fast Fashion *Fashion and Textiles* ***(3)25***
[28] Kobayashi H, Fukusige S 2018, A Living-Sphere Approach for Locally Oriented Sustainable Design, *Jnl Remanufactur*
[29] Handriana T, 2016 Mapping of Green Buying Perception in Developing Country *Mediterranean Journal of Social Sciences.MCSER Publishing Italy* ***(7)3****: 19 – 30*
[30] Shamsi M S, Siddiqui Z S 2017 Green Product and Consumer Behavior: An Analytical Study *Journal of Pertanika. Universiti Putra Malaysia Press*
[31] Ghazali I, Rashid S H A, Dawal S Z M, Tontowi A E, Aoyama H, Ghazilla R A R 2018 Green Product Preferences with Respect to Cultural Influences: Empirical Study in Indonesia *Int. J. Of Automation Technology* ***(12)6****: 842 – 852*
[32] Lim Y J, Perumal S, Ahmad N 2019 The Antecedents of Green Car Purchase Intention Among Malaysian Consumers *European Journal of Business and Management Research* ***(4)2***
[33] Chen C C, Chen C W, Tung Y C, 2018 Exploring the Consumer Behavior of Intention to Purchase Green Products in Belt and Road Countries: An Empirical Analysis *J. Sustainability* ***10***

[34] Lian S B, Yoong L C 2019 Assessing the Young Consumers' Motives and Purchase Behavior for Organic Food: An Empirical Evidence from A Developing Nation *International Journal of Academic Research in Business and Social Sciences* ***(9)1****: 69 – 87*

[35] Kowang T O, Samsudin S A, Yew L K, Hee O C, Fei G C, Long C S 2018 Factors Affecting Car Purchase Intenton Among Undergraduates in Malaysia *International Journal of Academic Research in Business and Social Sciences* ***8(8)****: 80 – 88*

[36] O'Rourke D, Ringer A 2015 The Impact of Sustainability Information on Consumer Decision Making *J. of Industrial Ecology*

[37] Hung N J, Yazdanifard R 2015 The Study of Vehicle Safety Aspects Influencing Malaysian Urban Consumer Car Purchasing Behavior *International Journal of Management Accounting and Economics* ***(2)8****: 913 – 924*

[38] Artsanti R, Soemarno, Yanuwiardi B, Djati M S 2018 Millenial's Preferences on Triple Bottom Line: Effect of The Corporate Social and Environmental Publicity on The Reputation, Attitude, and Intention to Work at a Company *Research on Humanities and Social Sciences* ***(8)18***

[39] Setyawan A, Noermijati N, Soenaryo S, Aisjah S 2018 Green Product Buying Intentions Among Young Consumers: Extending The Application of Theory of Planned Behavior *Problems and Perspectives in Management* ***(16)2****: 145 – 154*

[40] Michaud C, Llerena D 2010 Green Consumer Behavior: An Experimental Analysis of Willingness to Pay for Remanufactured Products *Business Strategy and The Environment Wiley Online Library*

[41] Cagalj M, Haas R, Morawetz U B 2016 Effects of Quality Claims on Willingness to Pay for Organic Foods *British Food Journal* ***118 (9)***

IOP Conf. Series: Materials Science and Engineering **857** (2020) 012025 doi:10.1088/1757-899X/857/1/012025

# Energetic and exergetic performance computation of coal-fired steam power plant using IAPWS IF-97 formulation for target variable datasets generation in machine learning based performance prediction

**A Y E Risano[1,*], A D Prabowo[1]**

[1] Department of Mechanical Engineering, Faculty of Engineering, Universitas Lampung, Gd. H Lt 2, Jalan Prof. Soemantri Brojonegoro No. 1, Bandar Lampung, 35143 Indonesia

[*] E-mail: ayerisano@gmail.com

**Abstract.** More than 50% of power plants in Indonesia are Coal-Fired Steam Power Plants. According to the Ministry of Energy and Mineral Resources of Indonesia, coal reserves in Indonesia will run out in 67 years. Fuel saving can be done by operating the system at the highest efficiency operating conditions. These conditions can be determined by analysing the operating history of the system along with the energetic and exergetic performance produced using machine learning algorithm. Unfortunately, energetic and exergetic performance calculation of the Coal-Fired Steam Power Plants system that is not easy results in lack of system performance target datasets generated from the history of the Coal-Fired Steam Power Plants operation, so that the Steam Powerplant performance data generation software written using the Python programming language is created in this paper to calculate the Coal-Fired Steam Power Plants energetic and exergetic performance using IAPWS IF-97 formulation quickly and accurately. Accuracy of the software written in this paper was tested using the Coal-Fired Steam Power Plants performance values calculated manually as a comparison and result difference between the two types of calculations below 1% and able to cut manual calculation time from 142.46 minutes to 14.34 minutes using the software which is feasible to generate target variable datasets needed for performance prediction using machine learning algorithm.

**Keywords:** coal-fired steam power plant, performance, IAPWS IF-97, machine learning

## 1. Introduction

More than 50% of the power plants in Indonesia are Coal-Fired Steam Power Plants. Coal is a non-renewable natural resource and according to the Ministry of Energy and Mineral Resources of Indonesia, coal reserves in Indonesia will run out in 67 years [1]. Therefore, if Indonesia is not ready to utilize other energy sources, then half of Indonesia will suffer a power outage. Using coal in an efficient way is needed to extend the time of coal fuel usage.

In order to commit coal fuel saving, operating coal-fired steam power plant at the highest efficiency condition is needed. Steam power plant performance can be determined analytically by the function of complete operating parameters datasets [2]. Based on Tarahan coal-fired steam power plant heat rate test report, some operating parameters such as LP turbine steam quality, turbine outlet steam quality, regenerator drain properties are not available and boiler emission properties are only available once a year so that analytical method will result low accuracy [3].

Another method to determine the highest efficiency operating condition is using machine learning algorithm with thousands of operating parameters datapoints. Researches show that this method results high accuracy with less operating parameter needed but required huge amount of the power generated target variable and operating parameter function variable datapoint as data feed [4,5]. In order to get more advanced result, more target variable such as turbine, boiler, condenser, pump, and regenerator performance based on energy balance and exergy are needed.

Adding these new target variables cannot avoid analytical calculation so that analytical and machine learning method need to be combined. Performance target and operating condition variable correlation can be determined without completing the data needed as analytic method but the performance target variable datasets still need to be calculated analytically. Unfortunately, these analytical calculations are not easy, consuming a lot of time and require special competencies that yield lack of datapoint. To solve the problem, a fast computation is needed to generate the variable target datasets corresponding to Coal-Fired Steam Power Plants PT PLN Persero Tarahan Generation Sector 100 MW which is the steam power plant object used in this paper.

## 2. Materials and Method

Coal-Fired Steam Power Plants PT PLN Persero Tarahan Generation Sector 100 MW is a type of regenerative Rankine cycle with 5 preheating levels through heat exchangers before feed water enters the boiler where the heat source comes from turbine extraction vapor. The specification has 23 main conditions [6] with the graph of the Rankine regenerative cycle as follows

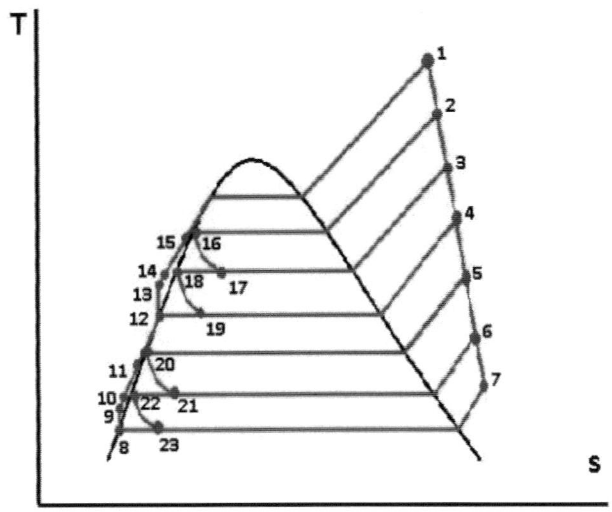


**Figure 1.** Tarahan Coal-Fired Steam Power Plants Rankine diagram

In order to calculate the performance of the main components of the Coal-Fired Steam Power Plants system, enthalpy and entropy values are needed which are functions of temperature and working fluid pressure. Manual enthalpy and entropy calculations can be done with the help of the Computer Aided Thermodynamics Table program as an alternative to the calculation by interpolating the values in the table of water-steam properties. While enthalpy and entropy calculations can automatically be done by creating an algorithm based on the IAPWS IF-97 standard which divides the water properties diagram into 5 areas [7] where each area has a different equation as shown below.

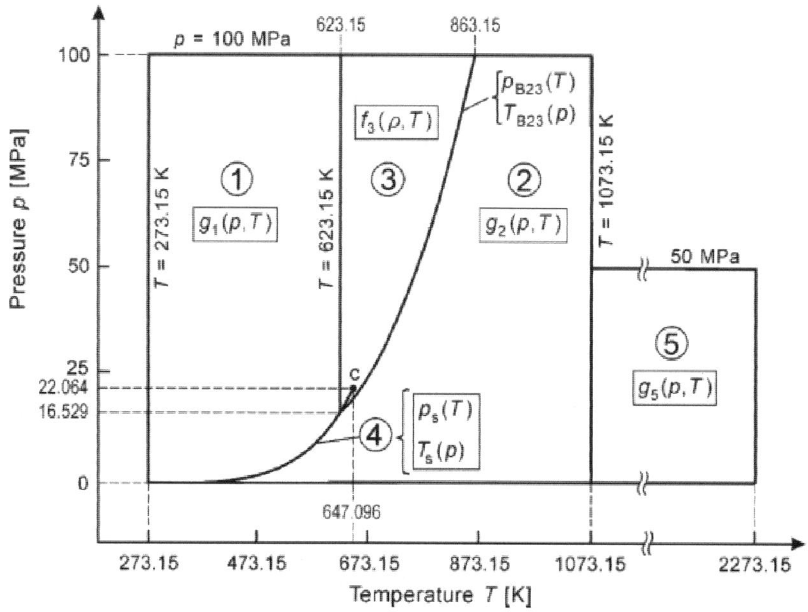


**Figure 2.** IAPWS IF-97 Diagram

Turbine performance can be obtained by calculating isentropic efficiency which is a comparison between actual work and isentropic work and exergetic efficiency which is a comparison of actual work with reversible work [8].

$$\eta_s = \frac{w_{aktual}}{w_s} \times 100\% \tag{1}$$

$$\eta_{II} = \frac{w_{aktual}}{w_{rev}} \times 100\% \tag{2}$$

Where:

$\eta_s$ : Isentropic efficiency (%)
$\eta_{II}$ : Exergetic efficiency (%)
$w_{aktual}$ : Actual work (kW)
$w_s$ : Isentropic work(kW)
$w_{rev}$ : Reversible work (kW)

Steam turbine performance calculation with 5 extraction points is divided into 6 segments based on the mass flow difference. this division occurs because the mass flow of steam through the turbine decreases due to the steam extraction, so it is necessary to calculate the amount of steam extracted at each extraction point using the principle of energy balance [8]. Pump performance is obtained by calculating isentropic efficiency and exergetic efficiency which is almost identical to the turbine but the actual work is in the lower section.

Boiler performance is obtained by boiler energetic efficiency calculation with the direct method or directly comparing the energy stored in the steam produced with energy stored in fuel and indirect methods that analyze the components that cause energy loss in the boiler [11]. Boiler performance is also calculated based on exergetic efficiency which is a comparison between exergy flow of steam with exergy of fuel and combustion air [10].

$$\eta_{II\ boiler} = \frac{\psi_{steam}}{\psi_{fuel}+\psi_{air}} \times 100 \tag{3}$$

Where:

$\eta_{II\ boiler}$ : Exergetic efficiency of boiler (%)
$\psi_{steam}$ : Steam exergy flow (kJ/kg)
$\psi_{fuel}$ : Fuel exergy flow (kJ/kg)
$\psi_{air}$ : Air exergy flow (kJ/kg)

The exergy of coal fuel is obtained by analyzing the chemical exergy of coal constituent substances using this following equation [9].

$$\psi_{ch} = \left[\left(NCV + 2442.\, m_{H_2O}\right)\varphi_{dry}\right] + 9417.\left(\frac{m_S}{100}\right) \tag{4}$$

Where:

$\psi_{ch}$ : Fuel exergy flow (kJ/kh)
$m_{H_2O}$ : water-fuel fraction
$\varphi_{dry}$ : dry fuel exergy (kJ/kh)
$m_s$ : Sulphur-fuel fraction
$NCV$ : Net Caloric Value fuel (kJ/kg)

The condenser and regenerator are heat exchangers. The performance of the components is obtained from the effectiveness calculation of heat exchanger. The effectiveness of a heat exchanger is a comparison between the actual operating heat load and the maximum heat load that can be transferred by a heat exchanger [12]. The objectives of this study is to create a software that is able to calculate energetic and exergetic performance of Coal-Fired Steam Power Plants PT PLN Persero Tarahan Generation Sector with a maximum value of 1% difference compared to manual calculation and significantly reduce calculation time to make combined analytical and machine learning method feasible to conduct.

## 3. Results and Discussions

### *3.1. Header and footers*

The software was written in Python 3 programming language with additional Scipy libraries to do scientific calculations, Pyqt5 library for building user interfaces, and iapws libraries to obtain enthalpy and entropy values from temperatures and pressures based on IAPWS IF-97 standards.

The program algorithm starts by calculating the enthalpy and entropy values from the temperature and pressure values entered based on the IAPWS IF-97 standard with corresponding filter created to minimize input error, then the enthalpy and entropy values generated are used as parameters in the Coal-Fired Steam Power Plants performance calculation as shown in the Fig. 3.

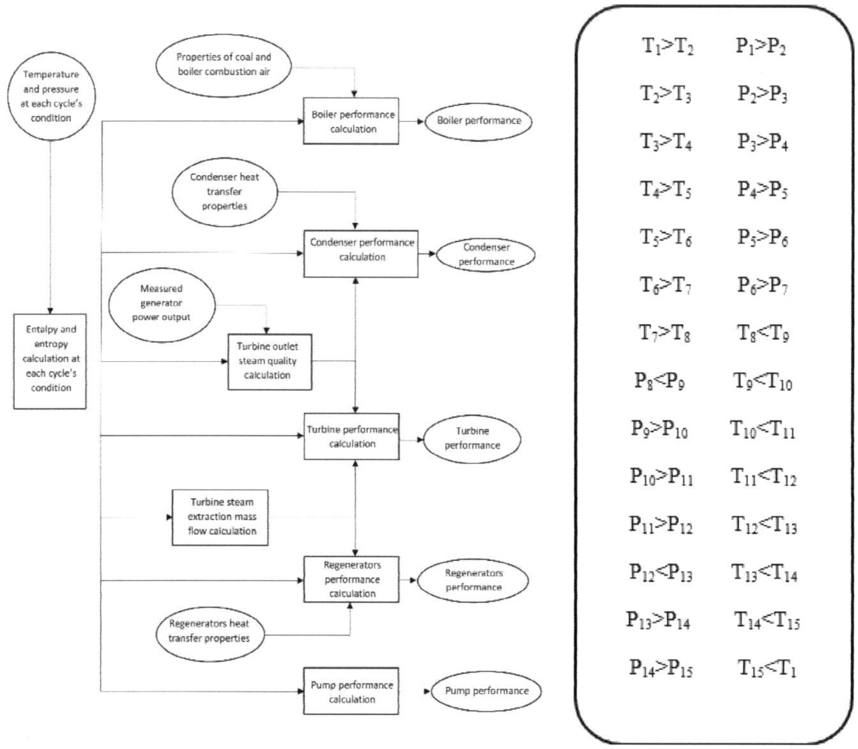


**Figure 3.** Software architecture and Input data filter

### *3.2. Validation and Software Testing*

Validation of the software output values performed by comparing the Coal-Fired Steam Power Plants performance values generated by the program with the performance values of manual calculation. The operation procedure of the software starts by entering the temperature, pressure, and mass flow values needed as well as the coal constituent substances as shown on Fig. 4.

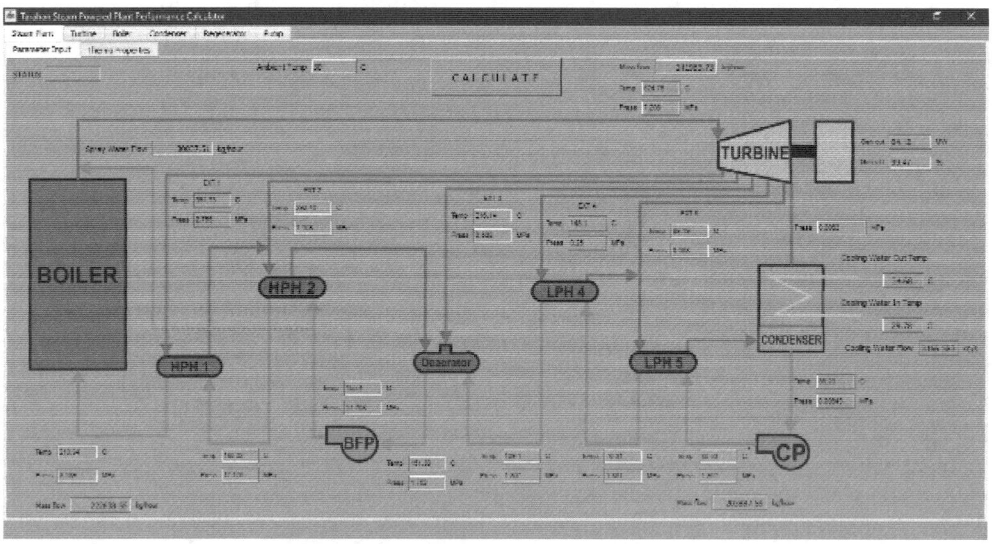


**Figure 4.** Input page of the software

IOP Conf. Series: Materials Science and Engineering **857** (2020) 012025 doi:10.1088/1757-899X/857/1/012025

The data used is the annual performance test results of PT PLN PERSERO Tarahan Generation Sector in 2017 with load variations of 64.12 MW, 74.25 MW, 86.02 MW, and 100.82 MW. As all the required data has been inputted, by clicking the calculate button, the calculation algorithm starts running and the performance value of the PLTU components will appear on each component's page. Re-checking the software input data is performed by looking at the status bar that is correct if the status "PASS" appears and somethings are incorrect if the "RECHECK" status appears.

The Coal-Fired Steam Power Plants main components performance values generated by the software are compared with the performance value obtained by manual calculations using the same operating data so that the difference between the two calculation methods are obtained as the following Table 1 to Table 6.

**Table 1.** Comparison of the calculation of steam turbine performance manually and using software.

| Efficiency | Load (MW) | Manual (%) | Software (%) | Difference (%) |
|---|---|---|---|---|
| Isentropic | 64.12 | 83.19 | 83.13 | 0.07 |
| | 74.25 | 86.63 | 86.61 | 0.02 |
| | 86.02 | 85.48 | 85.44 | 0.05 |
| | 100.82 | 82.54 | 82.43 | 0.13 |
| Exergetic | 64.12 | 84.17 | 84.15 | 0.02 |
| | 74.25 | 87.74 | 87.73 | 0.01 |
| | 86.02 | 86.69 | 86.66 | 0.03 |
| | 100.82 | 83.80 | 83.85 | 0.06 |

**Table 2**. Comparison of the calculation of boiler performance manually and using software

| Efficiency | Load (MW) | Manual (%) | Software (%) | Difference (%) |
|---|---|---|---|---|
| Energetic | 64.12 | 80.77 | 80.79 | 0.02 |
| | 74.25 | 75.70 | 75.72 | 0.03 |
| | 86.02 | 81.08 | 81.11 | 0.04 |
| | 100.82 | 86.78 | 86.78 | 0.00 |
| Exergetic | 64.12 | 26.85 | 26.86 | 0.04 |
| | 74.25 | 24.90 | 24.89 | 0.04 |
| | 86.02 | 26.21 | 26.22 | 0.04 |
| | 100.82 | 28.85 | 28.83 | 0.07 |

**Table 3**. Comparison of the calculation of condenser performance manually and using software

| Load (MW) | Effectiveness | | |
|---|---|---|---|
| | Manual (%) | Software (%) | Difference (%) |
| 64.12 | 54.94 | 54.92 | 0.03 |
| 74.25 | 59.84 | 59.81 | 0.04 |
| 86.02 | 58.63 | 58.62 | 0.02 |
| 100.82 | 54.53 | 54.52 | 0.02 |

**Table 4.** Comparison of the calculation of regenerators performance manually and using software.

| Regenerator | Load (MW) | Effectiveness | | |
|---|---|---|---|---|
| | | Manual (%) | Software (%) | Difference (%) |
| HPH 1 | 64.12 | 91.01 | 91.06 | 0.06 |
| | 74.25 | 86.63 | 86.67 | 0.05 |
| | 86.02 | 84.14 | 84.16 | 0.03 |
| | 100.82 | 78.99 | 79.03 | 0.05 |
| HPH 2 | 64.12 | 77.40 | 77.44 | 0.06 |
| | 74.25 | 73.97 | 74.08 | 0.14 |
| | 86.02 | 74.03 | 74.07 | 0.05 |
| | 100.82 | 70.40 | 70.46 | 0.08 |
| LPH 4 | 64.12 | 82.98 | 82.97 | 0.01 |
| | 74.25 | 77.36 | 77.36 | 0.00 |
| | 86.02 | 80.51 | 80.51 | 0.00 |
| | 100.82 | 78.29 | 78.32 | 0.04 |
| LPH 5 | 64.12 | 95.73 | 95.73 | 0.00 |
| | 74.25 | 92.19 | 91.95 | 0.26 |
| | 86.02 | 94.91 | 94.85 | 0.06 |
| | 100.82 | 93.61 | 93.53 | 0.08 |

**Table 5.** Comparison of the calculation of Condensate Pump (CP) performance manually and using software.

| Efficiency | Load (MW) | Manual (%) | Software (%) | Difference (%) |
|---|---|---|---|---|
| Isentropic | 64.12 | 64.75 | 64.88 | 0.02 |
| | 74.25 | 64.51 | 64.71 | 0.31 |
| | 86.02 | 69.17 | 69.16 | 0.01 |
| | 100.82 | 68.80 | 68.95 | 0.22 |
| Exergetic | 64.12 | 66.09 | 65.6 | 0.74 |
| | 74.25 | 65.86 | 65.57 | 0.44 |
| | 86.02 | 70.37 | 70.12 | 0.36 |
| | 100.82 | 69.86 | 69.77 | 0.13 |

**Table 6**. Comparison calculation of Boiler Feed Pump (BFP) performance (manually and software)

| Efficiency | Load (MW) | Manual (%) | Software (%) | Difference (%) |
|---|---|---|---|---|
| Isentropic | 64.12 | 63.89 | 63.99 | 0.16 |
| | 74.25 | 68.02 | 67.48 | 0.79 |
| | 86.02 | 86.26 | 86.26 | 0.00 |
| | 100.82 | 72.60 | 72.55 | 0.07 |
| Exergetic | 64.12 | 74.62 | 74.46 | 0.21 |
| | 74.25 | 77.68 | 77.39 | 0.37 |
| | 86.02 | 90.95 | 90.55 | 0.44 |
| | 100.82 | 81.09 | 81.05 | 0.05 |

Based on the data presented on Tables 1 to 6, there is no difference value that exceeds 1% between values obtained by manual calculation and generated by the software. Then the software operating speed testing was performed by 20 people with mechanical engineering background, each tested to calculate the energetic and exergetic performance manually aided with Computer Aided Thermodynamics Tables 3 and Microsoft excel filled with calculation template. Then the time consumed in manual calculation compared to the time consumed to calculate energetic and exergetic performance using software written. The result shows the software can have shortened energetic and exergetic performance calculation from 142.46 minutes manually to 14.34 minutes using the software.

## 4. Conclusion

A software that be able to calculate energetic and exergetic performance of Coal-Fired Steam Power Plants PT PLN Persero Tarahan Generation Sector was written. Performance value generated by the software and obtained by manual calculation was compared and found no difference value exceeds 1% and be able to shortened average calculation time from 142.46 minutes manually to 14.34 minutes using the software to generate 1 datapoints which is feasible to support combined analytical and machine learning method to conduct in future research

## Acknowledgment

The author acknowledges PT PLN PERSERO Tarahan Power Generation Sector for providing the field survey and the data required for the accuracy test of the software

## References

[1] Direktorat Jendral Mineral dan Batubara. 2016. “Laporan Kinerja 2016”. Jakarta: Kementrian Energi dan Sumber Daya Mineral Indonesia

[2] Mahmoud M, Abdella. 2019. “Parameters calculation of thermal power plant dynamic model using steam cycle data”. Elsevier: Thermal Science and Engineering Progress

[3] Pembangkit Listrik Negara. 2016. “Laporan pengujian heat rate PLTU Tarahan unit 3”. Lampung Selatan: PT PLN PERSERO Sektor Pembangkitan Tarahan.

[4] Ma, Teng. 2019. “Thermodynamic analysis and performance prediction on dynamic response characteristic of PCHE in 1000 MW S-CO2 coal fired power plant”. Elsevier: energy

[5] Pinar, Tufekci. 2014. “Prediction of full load electrical power output of a base load operated combined cycle power plant using machine learning methods”. Elsevier: International Journal of Electrical Power & Energy Systems.

[6] Subrata, Ari., Prabowo. 2015. “Analisa Termodinamika Pengaruh Aliran Massa Ekstraksi Turbin Uap Untuk Feedwater Heaters Terhadap Performa Pembangkit Listrik Tenaga Uap 100MW” Surabaya: Institut Teknologi Sepuluh Nopember

[7] Wagner, Wolfgang., Kretzschmar, Hans. 2008. “Properties of Water and Steam Based on Industrial Formulation IAPWS-IF97”. Berlin: Springer

[8] Yunus, A Cengel. 1998. “Thermodinamics: An engineering approach, 5th edition”. Boston: McGraw-Hill.

[9] Kaushik, S., Singh, Omendra. 2013. “Estimation of Chemical Exergy of Solid,Liquid, and Gaseous Fuels Used in Thermal Power Plants”. Budapest: Akademia Kiado

[10] Saidur, R., dkk. 2010. “Energy, Exergy, and Economic Analys of Industrial Boilers”. Elsevier: Journal of Energy Policy

[11] Burreau of Energy Efficiency. 2017.” Energy Performance Assesment of Boilers”. India: Ministry of Power India

[12] Frank, Incropera., David, Dewit. 2012. “Fundamental of Heat and Mass Transfer 7th Edition”. USA: John Wiley & Sons.

# Effect of porogen agent on microstructure of CaP granules using the gelation of alginate droplet formation

**M D Effendi[1*], D Gustiono[1], Lukmana[1]**

[1]Agency for The Assessment and Application of Technolgy, Gedung 224, Jl. Kw. Puspiptek, Muncul, Kec. Setu, Kota Tangerang Selatan, Banten 15314, Indonesia

*Email: dachyar.effendy@bppt.go.id

**Abstract**. CaP granules are used for open defect of bone. The important feature in the physical structure of a synthetic ceramic is its porosity. Pore structure is great importance for osteoconduction. The main objective of this study is to investigate the effect of adding carbon black as porogen agent on microstructure of CaP granules of droplet CaP. According to that, certain carbon black (325 mesh) has added into composition of microspheres CaP-alginate gelation with ratio CaP:Carbon black was 0.5, 1, and 1,5 weight ratios respectively. Scanning Electron Microscope (SEM) was used to investigate microstructure of droplet and shown that Irregular shape grain. Internal microstructure appears more open for samples produced without a presence of alginate after sintering process and confirmed by XRD pattern for whole samples. The porosity and pore size increased by increasing CB. On the other hand, no significant change on surface roughness and no transformation phase caused by adding CB. So that, it can be concluded that CB can serve as a porogen agent.

**Keywords**: porogent, CaP, granule, microstructure, porosity, droplet.

## 1. Introduction

Calcium phosphates have shown that these materials are suitable as bone substitutes due to their biocompatible, bioactive, biodegradable, and osteoconductive characteristics [1-3], and, when implanted in vivo, they are nontoxic and do not induce any antigenic response [4], so that, the calcium Phosphate can be applied as biomedicine [5] . The most widely used calcium phosphate ceramics are porous hydroxyapatite (HA), b-tricalcium phosphate (b-TCP), and mixtures of these two, known as biphasic calcium phosphate ceramics (BCP) [6]. The hydroxyapatite [Hap : $Ca_{10}(PO_4)_6(OH)_2$] is CaP, especially known to be excellent in biocompatibility because of its similarity in composition to natural bone [3][4]. For all of bone graft forms, it is important to make them spherically shaped for uniform packing in irregularly shaped defects [7][8].

The presence of porosity and a bioactive surface facilitate cell attachment, proliferation, and differentiation and, consequently, provide a more biocompatible, osteoconductive, and in some cases, osteoinductive ceramics, which can favor increased bone formation [9][10]. CaP scaffold or macroporous blocks have a more defined shape than granules and they possess big macro porosity (typical range 50-500 μm) that has a big role in osteoconductive mechanisms. It is impossible to achieve the same macro porosity in granules form, because the granules size is equivalent to macropores size.

Scaffolds have to be fixed into the open defect that is bigger than in the granules case and for the correct adaption to the implantation site they must to be drilled and cut for screw insertion. The different forms of granular calcium phosphate include irregular multifaceted granules and round smooth granules, with solid or porous structures. However, the behavior of granules in the body depends on their morphology and microstructure.

Bone structure is oriented according to physiological needs with interlinked tunnels, of non-uniform shape and width, whose boundaries are the trabeculae. The granulate model for bone simulation should also have microchannels (which in dentine and bone show identical extension of 1-2μm), which play an active role in cellular attraction processes on the surfaces [11]. Bioceramic micro porosity (pore size $<$ 10 μm), which is defined by its capacity to be impregnated by biological fluids, results from the sintering process, while the pore dimensions mainly depend on the material composition, thermal cycle and sintering time. Microporosity is an important parameter to achieve bioresorption at high osteogenic property; osteoconduction may be associated to the microporosity [12]. According to the importance of microporosity and surface properties as describe before, this work has investigated the effect of porogen agent on microstructure of CaP granules.

## 2. Materials and Methods

### *2.1 Materials*

*1) Materials preparation*

We used hydroxyapatite powder obtained from process as below:

Hydroxyapatite powder prepared by wet chemical methods of preparation 1M solution of calcium oxide and 0.6 gr/lt [$(NH_4)_2HPO_4$] is used as the initial solution. The solution of $(NH_4)_2HPO_4$ added drop-wise in a solution of calcium oxide. The level of addition of $(NH_4)_2HPO_4$varies for calcium precursors, then the influence of the size of the precursor at Hap morphology can be investigated. Ammonia was used for adjustment of pH values. The suspension is stirred with a magnetic stirrer after the addition of the initial solution for 24 hours at room temperature. Precursor done for 30 minutes at room temperature, then filtered and washed with a vacuum Buchner for five times until a neutral pH is reached. For the first and third phases were washed with distilled water and ammonia, the second phase and the fourth was washed with distilled water to remove residual dirt, and last washed with ethanol to remove water and increase the dispersibility. The end product was dried at room temperature for 24 hours then heated to $160^0$C for 17 hours for the removal of water. Sintering process was carried out at $650^0$C with a heating rate of 80 minutes with a time of 2 hours detention. Hydroxyapatite samples then characterized by XRD (Fig. 1)

The hydroxyapatite formed according to the reaction [13]:

$$10CaCO_3 + 5(NH_4)_2HPO_4 \rightarrow Ca_{10}(PO_4,CO_3)_6(OH,CO_3)_2 + 5(NH_4)_2CO_3 + 3CO_2 + 2H_2O. \quad 1)$$

*2) Droplet formation*

Droplet extrusion method allows the production of spherical drops starting from a bioceramic suspension containing Alginate solution that allows the instantaneous gelification in a crosslinking solution. Gelification mechanisms, as from previous studies [11-14], were applied to produce bioceramic spheres by droplet extrusion.

HA powder and certain carbon black ((CONTINEX N774, 325 mesh) has added into composition of microspheres CaP-alginate gelation with ratio CaP : carbon black were 0.5, 1, and 1,5 weight ratio respectively was mixed with sodium alginate solution(Sigma Aldrich ,3% w/v) at a ratio of 0.2 w/w and well homogenized. The paste was extruded drop-wise into a 0.1M $CaCl_2$ cross-linking solution, where spherical-shaped particles instantaneously formed and were allowed to harden for 30 min (fig. 1). The size of the microspheres was controlled by regulating the extrusion flow rate using a syringe. At completion of the gelling period, microspheres were recovered and rinsed in water in order to remove the excess $CaCl_2$. Finally, they were dried overnight in a vacuum-oven at 30°C, and then sintered at 1100°C for 1h, with a uniform heating rate of 5°C/min from room temperature (fig. 2). We made also CaP droplet without adding carbon black to compare between both samples' models, as shown in Fig. 3.

## 3. Result and Discussion

*3.1 Analysis of synthetic HA powder*

The wet chemical process as described on method to prepare hydroxyapatite powder successfully obtained hydroxyapatite as domain phase (Figure 1) and similar to carbonate- hydroxyapatite ($Ca_{10.00}P_{6.00}O_{26.14}H_{2.60}C_{0.02}$) pattern card number 96-900-3554.As we stated above, this reaction produced by product as ammonium carbonate and carbon dioxide, alongside water. The so-formed HAP is typically non-stoichiometric, as CO32− ions from CO2 in the atmosphere can partially replace PO43− and OH− ions, thus leading to formation of carbonated HAP, as stated in a review [2]. This result confirmed by Landi et al. 2003 with using the same process [15].

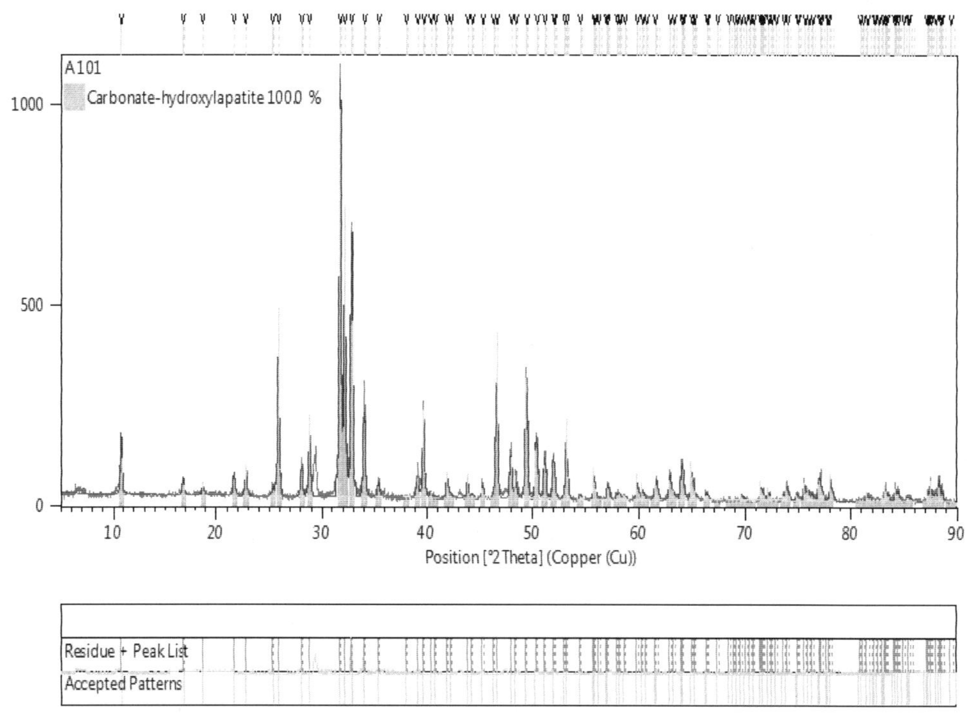


**Figure 1**. XRD pattern of hydroxyapatite powder obtained by wet chemical process and sintered at 650 °C.

*3.2 Analysis of HA granules*

Granules have fabricated by alginate gelation droplet formation. Carbon black was added into composition of microspheres HA-alginate gelation obtained black droplet as shown at figure2. Carbon black has completely burned after sintering process. It shown at figure.3 that droplets become white color compared to Fig. 2.

The pattern of XRD shown that all of samples had carbonate-hydroxyapatite ($Ca_{10.00}P_{6.00}O_{26.14}H_{2.60}C_{0.02}$) pattern identically as dominant phases. Effect of adding carbon black did not cause presence of any other phases on all ratio of CB added into the formulation. It also shown that after sintering process at $1000^0$ C for 1 hr holding time, CB and alginate had completely burned, and can be confirmed on FTIR investigation. The analysis using High Score Plus® shown that increasing CB added to granule formula caused lattice parameters slightly contracted (Table 1).

**Figure 2**. CaP granules/Carbon black before firing process(a), and after firing process (b) with the diameter are about 3mm

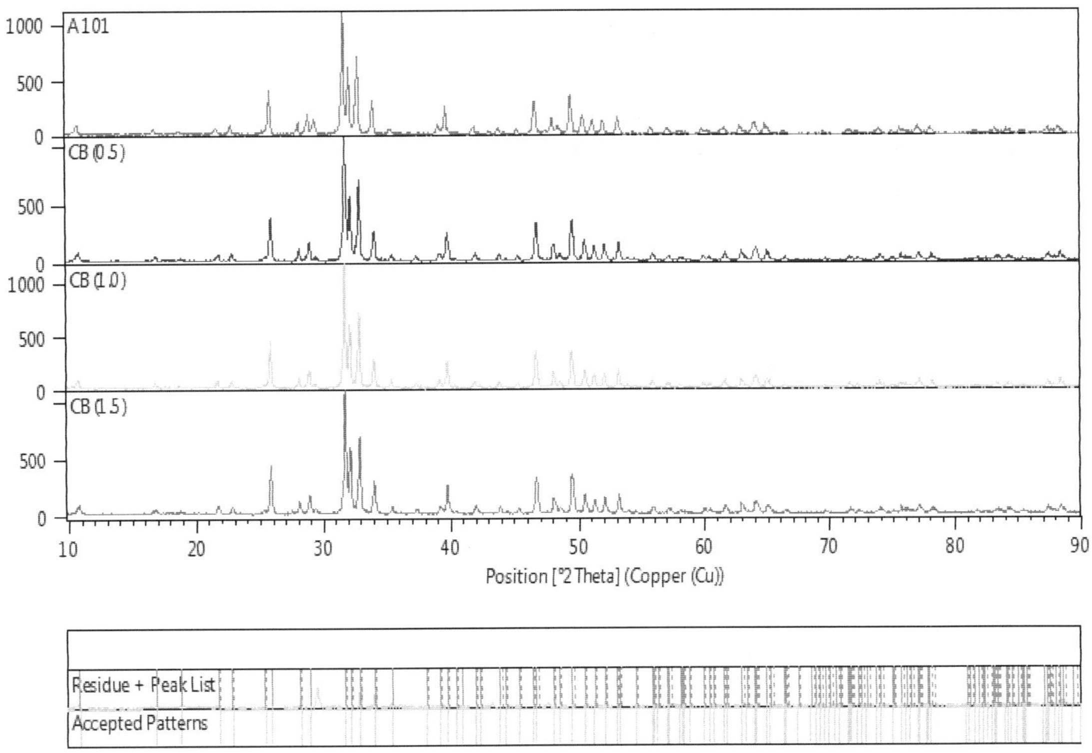


**Figure 3**. XRD pattern of granule HA adding 0.5,1.5,and 1.0 %wt respectively after sintering process at 1000 C , 2 hr.

**Table 1**. Structure and profile data of all samples based on XRD analysis using Highscore® software and Rietveld analysis

| Sample | Formula | Formula mass g/mol | Density (calc.) g/cm3 | Space group (No.) | Lattice parameters, Å | V/ $10^6$ $pm^3$ |
|---|---|---|---|---|---|---|
| **A101** | $Ca_{10.00}P_{6.00}O_{26.14}H_{2.60}C_{0.02}$ | 1007.79 | 3.16 | P63/m (176) | a :9.4292<br>b :9.4292<br>c :6.8736 | 529.2627 |
| **CB 0.5** | $Ca_{10.00}P_{6.00}O_{26.14}H_{2.60}C_{0.02}$ | 1007.79 | 3.16 | P63/m (176) | a :9.4240<br>b :9.4240<br>c :6.8784 | 529.0443 |
| **CB 1.0** | $Ca_{10.00}P_{6.00}O_{26.14}H_{2.60}C_{0.02}$ | 1007.79 | 3.17 | P 63/m (176) | a :9.4202<br>b :9.4202<br>c :6.8773 | 528.5235 |
| **CB 1.5** | $Ca_{10.00}P_{6.00}O_{26.14}H_{2.60}C_{0.02}$ | 1007.79 | 3.17 | P 63/m (176) | a :9.4189<br>b :9.4189<br>c :6.8771 | 528.3724 |

### *3.3 SEM analysis*

Granule for all samples type (with and without adding CB) shown that surface roughness, properties necessary for osteoconductivity (fig. 6-8). It could be conclusion that the more CB content, the more surface roughness founded. Irregular shape grain internal microstructure appears more open for samples produced without a presence of alginate after sintering process. The diameter of grain has decreased by adding CB and shown that, the effect of more containing CB (0.5% wt, 1.0 %wt, and 1.5% wt respectively has changed some grains into small particles with approx. 5- 10 um diameter. (fig 6-8). Consequently, the diameter of open porosity has decreased and microporosity increased.

The presence of macrostructural features, such as macropores, channels or void between particles has been shown to be a prerequisite for osteoinduction by synthetic biomaterials [15]. The macro porosity is not an intrinsic property of CaP granules, but represent the void space between granules. CaP ceramic microspheres can also use as drug delivery carriers, for example for doxycycline related to interconnected microporosity [16-18]. Thus, the presence of microposity should be an advantage to develop a granule which made by adding porogen to be a drug delivery carrier.

**Table 2**. Properties of the granules

| Sample Code | Surface roughness, μm | Pore size, μm | Porosity, % | Crystallite size |
|---|---|---|---|---|
| A101 | 0.25-2.1 | 0.12 - 0.40 | 51 | 359-762 |
| CB0.5 | 0.75-3.5 | 0.12 - 0.40 | 51 | 470-616 |
| CB1.0 | 0.6-1.8 | 0.13-0.40 | 53 | 454-759 |
| CB1.5 | 0.3-0.5 | 0.30 – 0.55 | 54 | 456-713 |

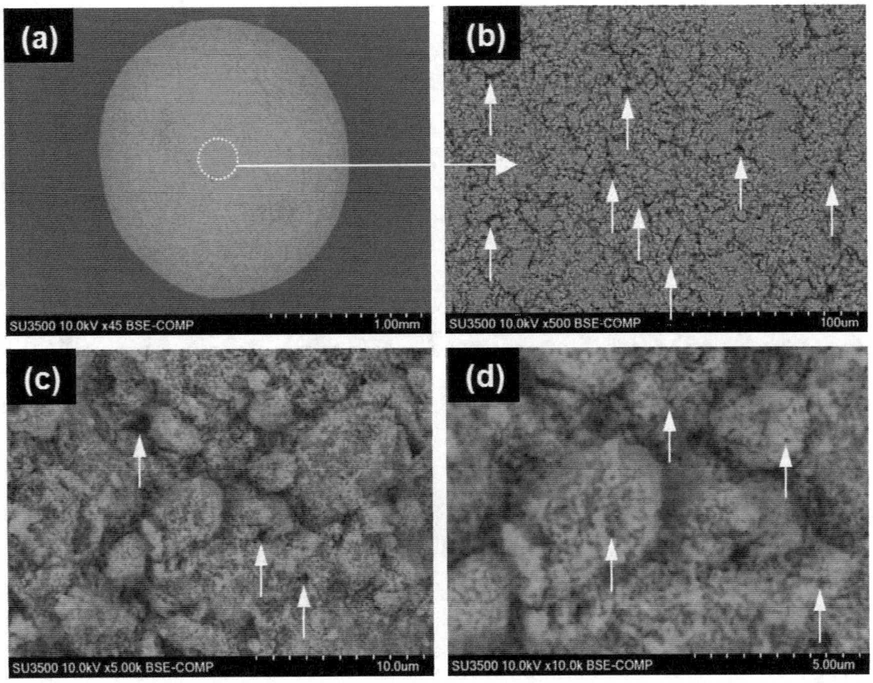


**Figure 4.** SEM micrograph of granules with adding 0.5% CB

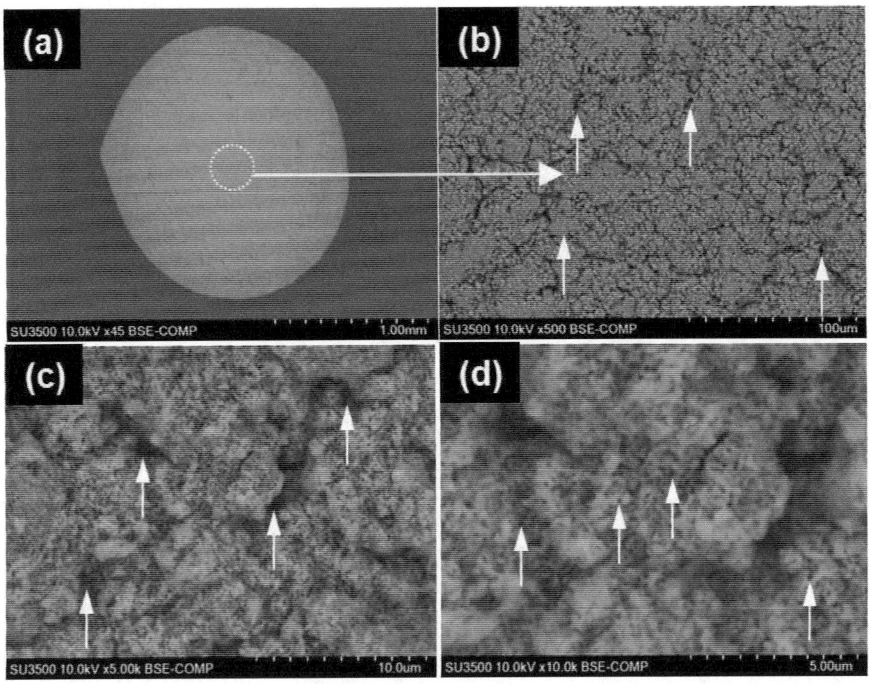


**Figure 5.** SEM micrograph of granules with adding 1% CB

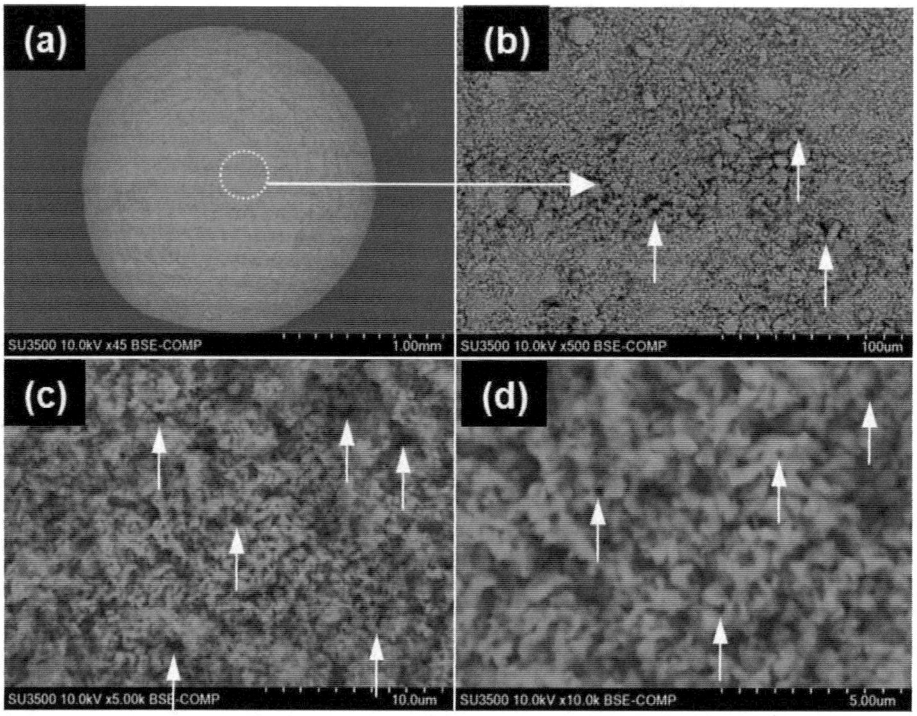


**Figure 6.** SEM micrograph of granules with adding 1.5% CB

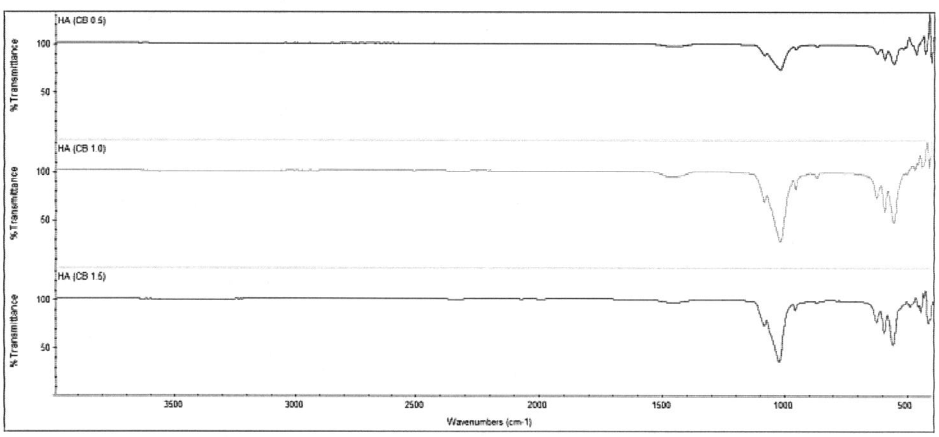


**Figure 7.** FTIR of granule HA adding 0.5 %wt, 1.0 % wt, and 1.5 %wt as sample representative of granule adding CB after sintering process at 1000$^0$C for 2hr

*3.4 FTIR analysis*

Infrared spectroscopy spectrum of HA granules shown in Fig 6. Strong bands attributed to PO43- groups. The bands at 1089 cm$^{-1}$ and 1026.03 cm$^{-1}$, 1088.49 cm$^{-1}$ and 1023.49 cm$^{-1}$, 1089.11 cm$^{-1}$ and 1027.45 cm$^{-1}$ band of CB0.5, CB1.0 and CB1.5 respectively are assigned to the components of the triply degenerated ν3 anti-symmetric P-O stretching mode. The 962.08 cm$^{-1}$, 962.10 cm$^{-1}$, and 962.34 cm$^{-1}$, band of CB0.5, CB1.0 and CB1.5 respectively is assigned to ν1. The non-degenerate P-O symmetric stretching mode. The bands at 600.31 $cm^{-1}$ and 564.30 $cm^{-1}$, 600.03 $cm^{-1}$ and563.34 $cm^{-1}$, and 600.58 $cm^{-1}$ and 564.87 $cm^{-1}$ bands of CB0.5, CB1.0 and CB1.5 respectively are assigned to components of the triply degenerate ν4 Q-P-O bending mode. Presence of any distinct band in the range of 1400-1550 $cm^{-1}$ indicates that the samples contain large quantities of carbonate ions. It means that hydroxyapatite as raw material has transformed into carbonated hydroxyapatite caused by presence of carbon ions. The broad band of low intensity in the range 3570-3580 $cm^{-1}$ coming from the hydroxyapatite OH stretching vibrations, corresponding to the hydroxyapatite.

## 4. Conclusion

A method for preparing granulates with varying pore size distribution from hydroxyapatite (HA) with adding carbon black as porogen can be made by droplet formation with help of porogen agent i.e carbon black according to this work. The shape and dimension of granules can be designed by adjusting diameter of syringe and depends in rheology of droplet granules. More porosity has been achieved with adding carbon black followed by increasing pore size. Producing hydroxyapatite by precipitation method produced carbonate hydroxyapatite. It is also can conclude that the porosity and pore size can be customized by adding carbon black on the formulation to produce HA granules by droplet extrusion. Granules can be designed for a wide range of medical applications

## References

[1] J. M. Ruan, J. P. Zou, J. N. Zhou, and J. Z. Hu. "Porous hydroxyapatite– tricalcium phosphate bioceramics," *Powder Metallurgy*, (2006), 49(1), pp. 66-69

[2] Habraken Wouter, Habibovic Pamela, Eppe Mathias, Bohner Marc."Calcium phosphates in biomedical applications: materials for the future?, Materialstoday,2016, 19(2), pp. 69-87

[3] H. Yuan, K. Kurashina, J. D. de Bruijn, Y. Li, K. de Groot, and X. Zhang, "A preliminary study on osteoinduction of two kinds of calcium phosphate ceramics," *Biomaterials*, vol. 20, no. 19, Oct. 1999, pp. 1799-806

[4] Jabr S. Al-Sanabani,1 Ahmed A. Madfa,2 and Fadhel A. Al-Sanabani, "Application of Calcium Phosphate Materials in Dentistry", International Journal of Biomaterials, 2013, vol. 2013

[5] M. Epple, K. Ganesan, R. Heumann et al., "Application of calcium phosphate nanoparticles in biomedicine," Journal of Materials Chemistry,2010, vol. 20, no. 1, pp. 18–23

[6] Landi E, Celotti G, Logroscino G, et al. "Carbonatedhydroxyapatite as bone substitute". J Eur Ceram Soc 25: 2003, 2931–2937.

[7] Ribeiro C.C., Barrias C.C., Barbosa M.A.: 'Preparation and characterisation of calcium-phosphate porous microspheres with a uniform size for biomedical applications', J. Mater. Sci. Mater. Med., 2006, 17(5), pp. 455–463

[8] Zhang, Ke et al. "Effect of microporosity on scaffolds for bone tissue engineering." Regenerative biomaterials, 2018, vol. 5(2): 115-124. doi:10.1093/rb/rby001

[9] Vallejos Baier R, Benjumeda Wijnhoven I, Irribarra Del Valle V, Millán Giovanetti C, Vivanco JF. "Microporosity Clustering Assessment in Calcium Phosphate Bioceramic Particles ". Front Bioeng Biotechnol. 2019;7: 281.2019.doi:10.3389/fbioe.2019.00281

[10] Suchanek W., Yoshimura M. 'Processing and properties of hydroxyapatite-based biomaterials for use as hard tissue replacement implants', J. Mater. Res., 1998, 13, (1), pp. 94–117.

[11] M. Fabbri, G. C. Celotti, and A. Ravaglioli, "Granulates based on calcium phosphate with controlled morphology and porosity for medical applications: physico-chemical parameters and production technique," *Biomaterials*, vol. 15, no. 6, May 1994, pp. 474-7.

[12] Musskaya, O.N., Kulak, A.I., Krut'ko, V.K. et al. "Preparation of Bioactive Mesoporous Calcium Phosphate Granules". Inorg Mater 54, 2018, 117–124 doi:10.1134/S0020168518020115

[13] Kamiya M., Hatta J., Shimada E., Ikuma Y., Yoshimura M., Monma H. AFM analysis of initial stage of reaction between calcite and phosphate. Mater. Sci. Eng. B Solid-State Mater. Adv. Technol. 2004; 111:226–231. doi: 10.1016/S0921-5107(04)00210-7.

[14] S. P. Victor and T. S. S. Kumar, "BCP ceramic microspheres as drug delivery carriers: synthesis, characterisation and doxycycline release,*Journal of materials science. Materials in medicine*, vol. 19, no. 1, Jan. 2008, pp. 283-90

[15] E. Landi, G. Celotti, G. Logroscino, and A. Tampieri, "Carbonated hydroxyapatite as bone substitute," *J. Eur. Ceram. Soc.*, 2003

[16] C. C. Ribeiro, C. C. Barrias, and M. a Barbosa, "Calcium phosphatealginate microspheres as enzyme delivery matrices," *Biomaterials*, vol. 25, no. 18, Aug. 2004, pp. 4363-73.

[17] P. L. Granja, a. I. N. Silva, J. P. Borges, C. C. Barrias, and I. F. Amaral,"Preparation and Characterization of Injectable ChitosanHydroxyapatite Microspheres," *Key Engineering Materials*, vol. 254-256, 2004, pp. 573-576.

[18] H. Yuan, H. Fernandes, P. Habibovic, J. D. Boer, A. M. C. Barradas, an A.D. Ruiter, "Osteoinductive ceramics as a synthetic alternative toautologous bone grafting," *Image (Rochester, N.Y.)*, vol. 107, no. 31, 2010, pp.10-15

# Implementation of fuzzy-profile matching in determining drug suitability for hypertensive patients

**A Wantoro[1,*], A Syarif[2], K Muludi[2], K Nisa[2]**

[1]Department of Computer Science, Teknokrat Indonesia University, Lampung, Indonesia
[2]Department of Computer Science, Faculty of Mathematic and Sciences, Lampung University, Bandar Lampung, Indonesia

*E-Mail: aguswantoro@teknokrat.ac.id

**Abstract.** Treatments of patients with hypertension can be done in two ways, namely modifying lifestyle and antihypertensive drugs. Generally giving hypertensive drugs depends on age, history of illness, smoking habits, obesity and on diseases such as diabetes, kidney failure, heart failure and ischemic heart disease. Inappropriate use of antihypertensive drugs often causes complaints such as fever, diarrhea, diabetes, mouth breathing, nerve disorders, radiation to the neck and head area. In providing hypertensive drugs that are appropriate to the patient's condition, doctor and medical personnels are required to have high pharmacological knowledge. Not all hospitals have pharmacology experts, therefore this study was conducted to develop a model that can be used as a drug suitability model based on the patient's condition. The method used in this study is Weight Matching Using Fuzzy Logic. Based on the results of comparisons with expert 1 and expert 2, the results of the average system testing are obtained with a value of 100% precision, recall 96.6% and accuracy 96.5%. Tests show that the application with a fuzzy logic approach can increase the efficiency of 9.5% compared to the interpolation weighting method.

**Keywords:** fuzzy logic, drug suitability, hypertensive patients, weight matching.

## 1. Introduction

Hypertension is a condition where blood pressure is >140/90 millimeters of mercury mmHG [1]. Hypertension is still the number one killer disease in Indonesia and hypertensive sufferers always experience an increase from 2013-2018 [2]. Hypertension is a very common medical condition, affecting roughly 20% of the world population [3]. Handling of patients with hypertension can be done in two ways, namely by modifying lifestyle and using antihypertensive drugs [4]. Anti-hypertensive drugs given usually are related to age, history of illness, smoking habits, obesity, and considering existing diseases such as diabetes, kidney, heart failure and ischemic heart [5].

The parameters used for the administration of hypertension drugs are age, weight, kidney health, lung health, sugar levels, blood circulation levels, heart health, depression levels, uric acid levels and the level of sexuality [5]. The use of anti-hypertensive drugs that are not appropriate can cause complaints or side effects such as fever, diarrhea, diabetes, kidney failure, stress, mouth breathing, nerve abnormalities, radiation to the neck and head area and local disorders of the salivary glands, therefore it is very important to know and know the type antihypertensive drugs [6]. Medications are the most common cause of reduced saliva. Antihypertensive drugs include one class of drugs that can cause side effects such as xerostomia. Data shows that around 770,000 patients in the hospital suffered injuries or even died from drug side effects. The primary source of error is prescribing errors [7].

Selection of the appropriate antihypertensive drug, the right dose, is a complex problem that requires a greater understanding of pharmacological options [8]. Based on data from the National Patient Safety Map Report 12$^{th}$ medication errors were ranked first at 24.8% of 10 cases. The results of an expert evaluation found a relationship between the types of hypertension drugs with the patient's health condition. There are 5 (five) types of hypertension drugs that have been identified, namely Diuretics, Beta Blockers, Ace Inhibitors, Calcium Channel Blockers (CCB) and Potential Receptor Blockers (ARBs) [5]. Even though various treatments have been carried out quite well, the level of uncontrolled hypertension remains high. Most patients need more than one drug because the recommendations in treatment are heterogeneous and no one has been able to identify optimal dosing strategies or drug combinations [9].

Research shows that 34% of anti-hypertensive drugs given by medical personnel are not the right dose [10]. In the United States, medication errors are estimated to harm 1.5 million patients per year. Not all hospitals have pharmacology experts. Therefore, this study was conducted to develop a model that can be used as a match for drugs based on the patient's condition. The study [5] conducted a drug evaluation using the Profile Matching method by weighting interpolation. The results showed an accuracy of 87%. This result is not optimal because of making inaccurate curves and domains. In weight parameters (BMI) if the patient weighs 13 bmi, the result of the value is 1.33 on a scale of 5. If calculated using an ideal table, then the membership value is 0. Therefore, this study will be developed using a fuzzy logic approach to calculate weighs to improve accuracy results.

The medical record data are used as input parameters of the patient's condition. Pharmacologist knowledge and drug parameters are used as input to determine the suitability of the drug using Profile Matching and Fuzzy Logic methods using curves and membership functions. The use of triangular and trapezoidal curves can provide smaller computational time [11]. The results of this study are in the form of a model of suitability of hypertension drugs with a better level of accuracy that can be used by doctors or medical personnel as a recommendation in providing medications for patients with hypertension.

## 2. The Proposed Method

The drug suitability model with the patient's health condition is developed by describing the proposed prototype architecture of the drug model. The prototype model consists of 2 (two) main parts, namely the development environment and consultation environment (Fig. 1).

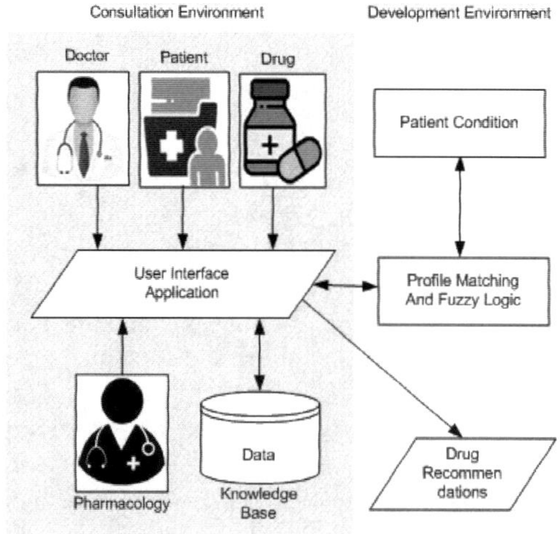


**Figure 1.** The suitability model of hypertension drugs is based on the patient's condition

### 2.1. Model Development Stages

The stages of development consist of creating knowledge base involving input data from hospitals, expert knowledge and knowledge from literature studies. The method used is Profile Matching method which is combined with fuzzy approach. Fuzzy logic is used to calculate the suitability of patients with the type of anti-hypertensive drug. The accuracy of the system is tested by experts using Confution Matrix table to get the values of accuracy, precision and recall, as presented on Fig. 2.

**Figure 2.** The stages of developing a conformity model for hypertension drugs

### 2.2. Knowledge Base Creation Stage

The stage of creating knowledge base is done by involving doctors/experts in the field of hypertension, considering the patient's condition data and antihypertensive drugs obtained from the patient's medical record, and the criteria used as a reference for the type of medicine that is appropriate to the patient's health condition.

### 2.3. Determination of Patient Parameters

The results of the consultation with the expert doctor indicated that there were 10 (ten) parameters that determined the administration of antihypertensive drugs in hypertensive patients. These criteria are age, weight, kidney health level, lung health level, sugar level, blood circulation level, heart health level, depression level, uric acid level and level of sexuality (table 1).

**Table 1.** Parameters of the patient condition that influence Hypertension drug administration[5]

| K1 | K2 | K3 | K4 | K5 | K6 | K7 | K8 | K9 | K10 |
|---|---|---|---|---|---|---|---|---|---|
| Age | Weight | health kidney level | Health lung level | Grade sugar level | Circulation level blood | Heart health level | Level depression | Uric acid levels (K9) | Sexuality level |
| year | BMI | mg/dl | ppm | mg/dl | liter | pulse/minute | Hamilton | mg/dl | ordinal |

### 2.4. Suitabilty of Condition of Patients with Types of Hypertension Medication

To find out the suitability of the patient's condition with the type of Hypertension drug, it is necessary to have ideal data from 10 parameters with the suitability of certain types of hypertensive drugs, as well as other types of drugs such as those in show in table 2.

**Table 2.** Value of ideal criteria with types of antihypertensive drugs [5]

| Type of drug | K1 | K2 | K3 | K4 | K5 | K6 | K7 | K8 | K9 | K10 |
|---|---|---|---|---|---|---|---|---|---|---|
| **Diuretic** | ≤ 65 | 18,5-24,9 | 0,6-1,2 | < 4 | 60-139 | > 6 | > 100 | ≥ 35 | 2-7,5 | Up Normal |
| **Beta B** | ≤ 65 | 18,5-24,9 | > 1,2 | < 4 | 60-139 | > 6 | 60-100 | ≤ 35 | > 7,5 | Normal |
| **ACE-I** | > 65 | ≥ 25 | > 1,2 | ≥ 4 | ≥ 140 | > 6 | > 100 | > 35 | > 7,5 | Normal |
| **CCB** | ≤ 65 | ≥ 25 | > 1,2 | ≥ 4 | ≥ 200 | 5 – 6 | > 100 | ≥ 35 | > 7,5 | Up Normal |
| **ARB** | ≤ 65 | ≥ 25 | 0,6-1,2 | ≥ 4 | ≥ 200 | > 6 | > 100 | ≥ 35 | > 7,5 | Up Normal |

Furthermore, this table is used as an ideal value to determine the weight value of each parameter that has matches the condition of the patient with the type of hypertension drug using Matching and Fuzzy logic methods.

### *2.5. Making Ideal Membership Curves and Functions of Drug Types*

Forming curves based on patient parameters for each type of drug is carried out based on the ideal profile value of each parameter that has been verified by a specialist. Giving amounts formulated using fuzzy logic, as presented on table 3.

**Table 3.** Curves and membership functions are ideal for types of Diuretic drugs

| Parameter | Curve | Membership Function |
|---|---|---|
| Age (K1) | 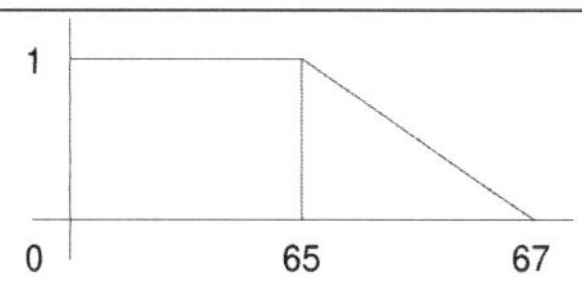  | $x[K] = \begin{cases} 1 & x \le 65 \\ \frac{65-x}{65-60} & 65 < x \le 67 \\ 0 & x > 67 \end{cases}$ |
| Weight (K2) | 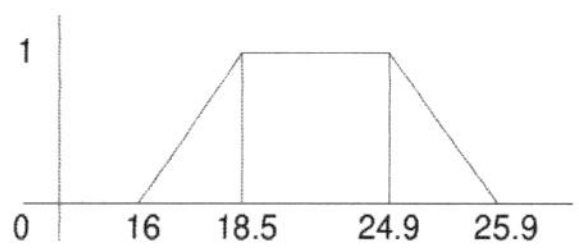  | $x[K] = \begin{cases} \frac{x-16}{18.5-16} & 16 \le x \le 18.5 \\ 1 & 18.5 \le x \le 24.9 \\ \frac{25-x}{25.9-24.9} & 24.9 < x \le 25.9 \\ 0 & 16 < x > 25.9 \end{cases}$ |
| Health Kidney level (K3) | 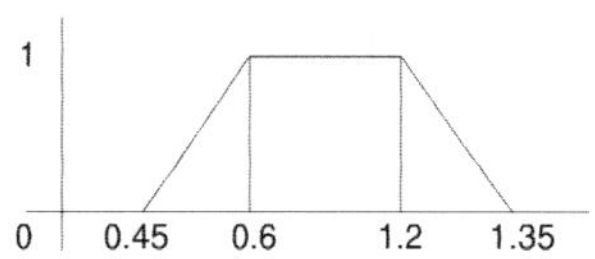  | $x[K] = \begin{cases} \frac{x-0.45}{0.6-0.45} & 0.45 \le x \le 0.6 \\ 1 & 0.6 \le x \le 1.2 \\ \frac{1.4-x}{1.4-1.2} & 1.2 < x \le 1.35 \\ 0 & 0.45 < x > 1.35 \end{cases}$ |
| Health lung level (K4) | 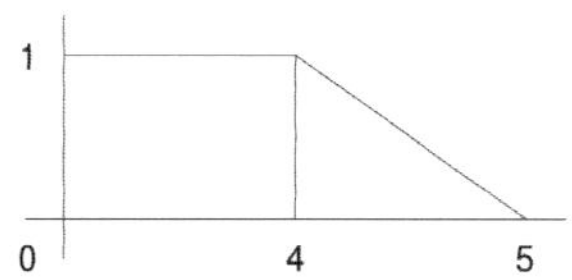  | $x[K] = \begin{cases} 1 & x \le 4 \\ \frac{5-x}{5-4} & 4 \le x \le 5 \\ 0 & x > 5 \end{cases}$ |
| Level of sugar (K5) | 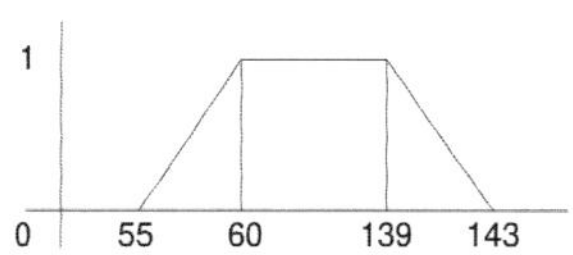  | $x[K] = \begin{cases} \frac{x-55}{60-55} & 55 \le x \le 60 \\ 1 & 60 \le x < 139 \\ \frac{143-x}{143-139} & 139 < x \le 143 \\ 0 & 55 < x > 143 \end{cases}$ |
| Circulation blood level (K6) | 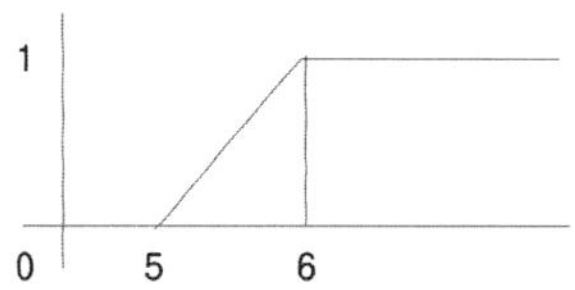  | $x[K] = \begin{cases} 0 & x < 5 \\ \frac{6-x}{6-5} & 5 \le x \le 6 \\ 1 & x > 6 \end{cases}$ |
| Health heart level (K7) | 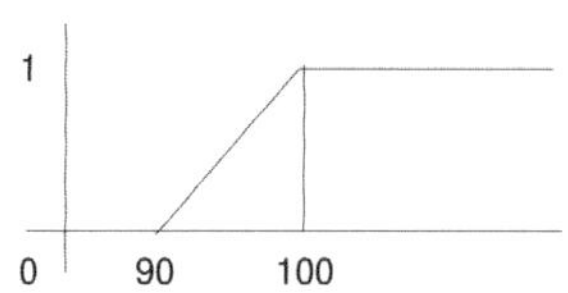  | $x[K] = \begin{cases} 0 & x < 90 \\ \frac{x-90}{100-90} & 90 \le x \le 100 \\ 1 & x > 100 \end{cases}$ |

| | | |
|---|---|---|
| Depression level (K8) | 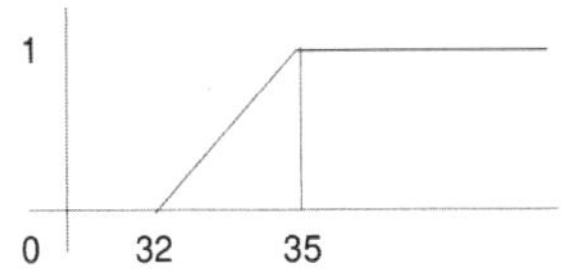 | $x\,[K] = \begin{cases} 0 & x < 32 \\ \dfrac{x-32}{35-32} & 32 \le x < 35 \\ 1 & x \ge 35 \end{cases}$ |
| Uric acid levels (K9) | 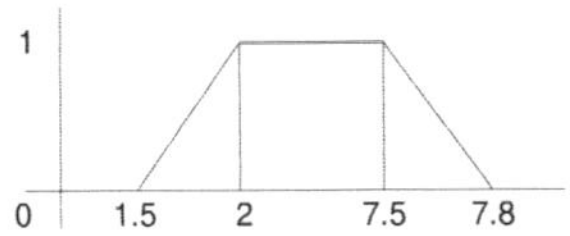 | $x\,[K] = \begin{cases} \dfrac{x-1.5}{2-1.5} & 1.5 \le x \le 2 \\ 1 & 2 \le x < 7.5 \\ \dfrac{7.8-x}{7.8-7.5} & 7.5 < x \le 8 \\ 0 & 1.5 < x > 8 \end{cases}$ |
| Sexuality (K10) | - | $x\,[K] = \begin{cases} 0 & Low\ normal \\ 0.5 & Normal \\ 1 & Up\ Normal \end{cases}$ |

Also, the ideal curves for Beta Blocker type drugs, Ace Inhibitors, CCB and ARB is performed as on table 4 and 5, where the calculation of membership weights for Patient "P-001" on the suitability of the Diuretic drug.

**Table 4.** Calculation of the value of W (x) of the patient "P-001" with the Type of Diuretic Drug

| P-001 | Data | Membership Function |
|---|---|---|
| Age | 76 | W(x) = (>67) = 0 |
| Weight | 17 | W(x) = (17-16)/(18.5-16) = 0.4 |
| Health kidney level | 1.5 | W(x) = (>1.35) = 0 |
| Health Lung level | 7.3 | W(x) = (>5) = 0 |
| Sugar level | 95 | W(x) = (60<= x <=139) = 1 |
| Circulation blood level | 11 | W(x) = (x>6) = 1 |
| Health heart level | 114 | W(x) = (x>100) = 1 |
| Depression level | 19 | W(x) = (x<32) = 0 |
| Uric acid levels | 6 | W(x) = (2<=x<=7.5) = 1 |
| Sexuality | Low normal | W(x) = 0 |

**Table 5.** Calculation of weights for types of Diuretic drugs

| Patient | K1 | K2 | K3 | K4 | K5 | K6 | K7 | K8 | K9 | K10 |
|---|---|---|---|---|---|---|---|---|---|---|
| P-001 | 76 | 17 | 1,5 | 7,3 | 95 | 11 | 114 | 19 | 6 | Up normal |
| Value | 0 | 0.4 | 0 | 0 | 1 | 1 | 1 | 0 | 1 | 0 |

### *2.6. Determination of Core Factor (NCF) and Secondary Factor (NSF) Value*

The next stage is grouping patient parameters into core factors and secondary factors. Core factors are the leading parameter group where the determination of the type of drug given is very dependent on the parameters in this group while the secondary factor is a parameter group that does not have a strong influence on the determination of the type of drug given to hypertensive patients (table 6).

**Table 6.** Parameters of core factors and secondary factors

| *Core Factor (CF)* | *Secondary Factor (SF)* |
|---|---|
| *1. Kidney health level (K3)* | *1. Age (K1)* |
| *2. Level of sugar (K5)* | *2. Weight (K2)* |
| *3. Blood circulation level (K6)* | *3. Lung health level (K4)* |
| *4. Heart health level (K7)* | *4. Depression level (K8)* |
| *5. Uric acid level (K9)* | *5. Sexuality (K10)* |

The calculation of the weight value is regarded as the average weight value of the core factor parameters (level of kidney health, sugar level, arterial blood circulation level, heart health level, level of uric acid level) and secondary core factors (age, weight, level of lung health, Level of depression, sexuality) as below:

Average weight of core factor parameters (NCF)

$$NCF = \frac{(0+1+1+1+1)}{5} = 0.8$$

Average weight of secondary factor parameters (NSF)

$$NSF = \frac{(0+0.4+0+0+0)}{5} = 0.08$$

The value of suitability evaluation of patients with diuretic drugs was calculated by taking into account the influence of core factors and secondary factors as follows:

$$\begin{aligned} \text{Match value} &= (0.60 * NCF) + (0.4 * NSF) \\ &= (0.60 * 0.8) + (0.4 * 0.08) \\ &= 0.48 + 0.032 \\ &= 0.512 \end{aligned}$$

The result value of 0.512 indicates that patient "P-001" with the conditions shown in the patient data table, if given a type of Diuretic anti-hypertension drug, shows a match (0.512 / 1) x 100% = 51.2%. By using the same method the match value of patient "P-001" with other types of drugs can be obtained (table 7).

**Table 7.** Results of patient match rates for each type of antihypertensive drug

| Patient | Diuretic | Beta B | ACE-i | CCB | ARB |
|---|---|---|---|---|---|
| P-001 | 51.2% | 47.2% | 52% | 32% | 32% |

Thus, the model can calculate the suitability of the patient's condition with the type of antihypertensive drug with different drug matches.

### *2.7. Determination of recommendations for drug administration*

The total value of the calculation of the suitability of the type of drug and the patient's health parameters are then sorted from the largest to the smallest. The type of drug with the first order of suitability or the largest total value is recommended to be given to the patient because the drug is best suited to the patient's health condition. However, all calculation results are displayed so that doctors and medical personnel can determine based on their level of expertise and experience.

## 3. Testing and Results

The test conducted includes calculating the value of accuracy, precision and recall to show the value of accuracy. Tests were carried out in 2 (two) stages involving different experts. Each experiment used the same 100 patient data in previous studies. The test results produced an average value of 100% precision, 94.7% recall and 95% accuracy (table 8).

**Table 8.** Confusion matrix with the first expert

| | Type of Drug | ARB | ACE | Beta | CCB | Diuretic |
|---|---|---|---|---|---|---|
| Actual Table | **ARB** | 6 | 0 | 0 | 0 | 0 |
| | **ACE** | 0 | 28 | 0 | 0 | 1 |
| | **Beta** | 0 | 0 | 29 | 0 | 1 |
| | **CCB** | 0 | 0 | 0 | 6 | 1 |
| | **Diuretic** | 0 | 1 | 0 | 1 | 31 |

The second test conducted by different experts showed better accuracy with 100% precision, 98.5 recall and 98% accuracy. This means that the proposed model can provide good drug recommendations (table 9).

**Table 9.** Confusion matrix with the second expert

| | Type of Drug | ARB | ACE | Beta | CCB | Diuretic |
|---|---|---|---|---|---|---|
| Actual Table | **ARB** | 6 | 0 | 0 | 0 | 0 |
| | **ACE** | 0 | 28 | 0 | 0 | 0 |
| | **Beta** | 0 | 0 | 28 | 0 | 1 |
| | **CCB** | 0 | 0 | 0 | 5 | 0 |
| | **Diuretic** | 0 | 0 | 1 | 0 | 31 |

Based on the values of accuracy, precision and recall in table 8 and table 9, the average value is calculated with a value of 100% precision, 96.6% recall and 96.5% accuracy. This value indicates that fuzzy logic approach can increase the value of accuracy by 9.5% when compared to interpolation approach. The system can provide recommendations for drug administration correctly.

### *3.1. Comparison of Fuzzy Weighting, Interpolation and Ordinal*

In the previous studies, an ordinal weighting and interpolation method were proposed, so to see the accuracy with fuzzy approach a comparison with ordinal methods and interpolation was needed. Figure 3 presents the results of a comparison of the values of accuracy, precision and recall of the three (3) methods (Fig. 3).

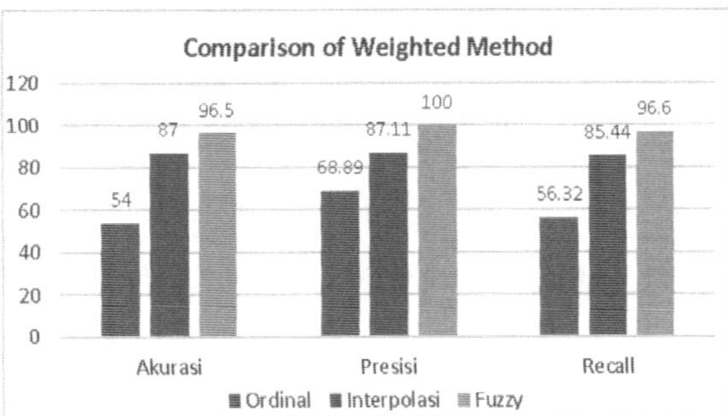


**Figure 3.** Comparison of weighted *Ordinal*, *Interpolation* and *Fuzzy*

## 4. Conclusion

Based on the description, explanation and testing that have been carried out, some conclusions have been drawn that this study produced a model of suitability of drug administration based on the patient's health conditions using Profile Matching and fuzzy logic approach. Based on the results of comparisons with expert 1 and expert 2, the results of the average system testing are obtained with a value of 100% precision, recall 96.6% and accuracy 95%. Tests show that the application with fuzzy logic approach can increase the accuracy of 9% compared to the interpolation weighting method. However, this research still needs to be continued, considering that it still has several weaknesses and disadvantages. Creating knowledge base to be continually updated and adapted to the patient's condition for different sexes is essential.

**Acknowledgments**
The author would like to thank the University of Teknokrat Indonesia who has provided support, to the Promoter and Co-Promoter for his guidance and to the IC-STAR 2019 Committee thank you for giving this opportunity.

**References**

[1] A. WHO., "Hypertension fact sheet," *Hypertension*, pp. 1–2, 2011.
[2] Riset Kesehatan Dasar, "Hasil Utama Riskesdas 2018," *Kementrian Kesehat. Republik Indones.*, 2018.
[3] WHO, "Global brief on hypertension," *World Heal. Organ. Geneva. WHO/DCO/WHD/2013.2*, pp. 1–40, 2013.
[4] Indonesia Ministry of Health, "Pusdatin Hipertensi," *Infodatin*, no. Hypertension, pp. 1–7, 2014.
[5] H. Soetanto, S. Hartati, R. Wardoyo, and S. Wibowo, "Hypertension drug suitability evaluation based on patient condition with improved profile matching," *Indones. J. Electr. Eng. Comput. Sci.*, 2018.
[6] C. Fox, P. L. Kalarickal, M. J. Yarborough, and J. Y. Jin, "Perioperative management including new pharmacological vistas for patients with pulmonary hypertension for noncardiac surgery," *Current Opinion in Anaesthesiology*. 2008.
[7] R. Koppel *et al.*, "Role of Computerized," *J. Am. Med. Assoc.*, vol. 293, no. 10, pp. 1197–1203, 2005.
[8] F. W. Germino, "The Management and Treatment of Hypertension," *Clin. Cornerstone*, vol. 9, no. SUPPL. 3, pp. 27–33, 2008.
[9] G. Koren, G. Nordon, K. Radinsky, and V. Shalev, "Machine learning of big data in gaining insight into successful treatment of hypertension," *Pharmacol. Res. Perspect.*, vol. 6, no. 3, pp. 1–6, 2018.
[10] W. Supadmi, "Evaluation of Drug Use in Anti Hypertension Patients Who Chronic Renal Failure Undergo Hemodialysis," *J. Ilm. Kefarmasian*, 2011.
[11] P. Baranyi, T. D. Gedeon, and L. T. Koczy, "A general interpolation technique in fuzzy rule bases with arbitrary membership functions," pp. 510–515, 2002.

# Deep-hair-cracks mechanism of rigid pavement in humid tropical weather

**C Niken[1*], Rainal[1], M Karami[1], and P Sasana[1]**

[1] Department of Civil Engineering, Universitas Lampung, Jalan Prof. Soemantri Brojonegoro No.1, Bandar Lampung 35145, Indonesia

*Email: chatarinaniken@yahoo.com

**Abstract.** This research aims to find the cause of deep hair crack in the rigid pavement. This research was done by the observation on a real rigid pavement in Lampung, Indonesia. Rigid pavement length is 39.4 km with four lanes. The length was divided into segments, in which every segment is 5m × 4.6m. The total segments are 31520. The thickness of the rigid pavement is 30 cm with a flexural strength of 45 MPa. The base of the rigid pavement is lean concrete with 10 cm depth and fc' of 10 MPa. The rigid pavement has contacted with the environment during the first 2 hours because of the made of grooving and cutting. The hair-cracks appear at 2 hours of age on 231 segments or 0.1% of all observed segments. Three core drills were done at the rigid pavement with hair-cracks at the age of 8 months. The depth of the cracks reached 77% of the rigid pavement depth. Hair-cracks appear on the surface because CH bond is not strong and elastic enough to withstand the tensile stress on the surface, the change of high rate expansion and high rate shrinkage due to $Ca^{2+}$.

**Keywords:** Concrete, cracks, pavement, rigid pavement, humid tropical weather

## 1. Introduction

Minor deformities or imperfections which are limited to the surface of concrete pavement are often referred to as surface defects. All of the surface defects can be affected by changing weather conditions, unique mix design characteristics, and de-icing/ anti-icing practice. These distresses typically do not significantly detract from the structural integrity of the pavement but can have an impact on its functional performance and its aesthetic appeal. These surface defects include the following: map cracking or crazing, plastic shrinkage, scaling, surface polishing, surface wear, and mortar flaking [1]. Concrete structures do not frequently fail due to lack of strength, but due to inadequate durability or due to improper maintenance technique. Despite excellent service history concrete pavement faces accelerated deterioration due to water and chemical ingress through the micro crack and surface void [2].

The quality of concrete floor or slab is highly dependent on achieving a hard and durable surface that is flat, relatively free of cracks, and at the proper grade and elevation. Properties of the surface are determined by the mixture proportions and the quality of the concreting and jointing operations, especially finishing, jointing, and curing is critical. Failure to address this issue can contribute to undesirable characteristics in the wearing surface such as cracking, low resistance to wear, dusting, scaling, high or low spots, poor drainage, and increasing the potential for curling. Concrete floor slabs employing Portland cement, regardless of slump, will start to experience a reduction in volume as soon as they placed. This phenomenon will continue as long as any water, heat, or both is being released to the surrounding [3]. The hairline cracks occur more readily for the finer grade of sand than in the case of the coarser grade of sand [4].

Hair cracks type is random and the width is small. Map or random cracking is a network of cracks and fine fissures on the concrete surface that enclose small and irregularly shaped areas which size dimensions less than 2 in [5]. If larger cracks occur, are frequently oriented in the longitudinal direction of the pavement and are interconnected by finer transverse or random cracks. These cracks are shallow, often only 3 mm deep. Map cracking may be localized or may occur over the entire surface of the concrete slab. The number of occurrences and affected areas (in $ft^2$ or $m^2$) is recorded. When an entire section is influenced, it is considered a single occurrence.

The most common cause of premature deterioration is attributed to the development of cracks. Cracking can occur in concrete pavements and structures for several reasons that can primarily load or environmental effects. Within the first few hours after placement, plastic shrinkage can occur at the surface of fresh concrete. Plastic cracks typically do not exceed 10mm wide but may pass through the full depth of the member. The mechanism leading to the formation of plastic shrinkage cracking does not explain full depth cracks. Plastic shrinkage cracking is a problem for large flat structures in which the exposed surface area is high relative to the volume of the placed concrete. The crack caused by plastic shrinkage can be quite wide on the upper surface 2 to 3 mm, but their width often decreases rapidly below the surface. It is probable that the subsequence events, including drying shrinkage and loading, can cause the plastic shrinkage cracks to propagate. Shrinkage is the main factor in the development of these cracks. Random, pattern, or map cracking were caused by plastic shrinkage and drying shrinkage. The time of appearance is 30 min to 6 hours and weeks to months, respectively. The environment has a significant influence on the properties of fresh and hardened concrete. Relative humidity, temperature, and wind show that the construction environment is a major concern when assessing the risk of this early age cracking [5].

The response of materials depends on environmental conditions. The air temperature and relative humidity are changing time by time; therefore, its influence on the rigid pavement is not stable. The temperature effect is usually considered to be less important than relative humidity [6]. Although, it is less critical, at the early age ambient temperature is more dominant than humidity because of the ambient temperature in the form of gas with many particles with high-temperature moves quickly and penetrates the rigid pavement easily. Indonesia is the area with high rainfall. The crack observation is in Oslo [7]. The temperature range in Oslo is about -7-22°C, humidity 64-81%, and rainfall occur over a year. Pouring under hot weather even can lead to conditions that can contribute to rapid moisture loss and the development of plastic shrinkage cracking. Rapidly changing weather conditions can affect the quality of the curing. When it happens, the durability and strength of the concrete surface are also affected [8]. Highly evaporative conditions, e.g., windy conditions, and exposure to direct sunlight, contribute to rapid moisture loss from the surface, which can lead to the development of plastic shrinkage cracking in the concrete. This condition makes the mechanism of deep-hair-cracking different in the area with four seasons, and humid tropical weather, moreover, coupled with high rainfall conditions. By knowing the cause, preventive action can be done rightly.

## 2. Methods

The research was done by observed hair-crack of a rigid pavement (RP) in National road Bakauheni-Sidomulyo, Lampung Province, Indonesia. RP length observed is 39.4 km with four lanes. The closest distance to the highway to the beach is 10 km. The length was divided into segments. The measure of every segment is 5m × 4.6m. The total segments are 31520 pieces.

### *2.1. Materials*

The RP was placed on 10 cm thick of lean-concrete (LC) with a compressive strength of 10 MPa. The cross-section layer of the RP can be shown in Figure 1.

**Figure 1**. Cross-section layer of the rigid pavement

Pouring was begun at about 3 pm to decrease ambient temperature effect to the concrete. After pouring, RP does not cover the surface until groove making. Curing compounds should be done right after grooving to maintain internal moisture. It was done about a half-hour after grooving. Curing compound Type 2 according to [9] were used. The curing compound will disappear after about 5-8 hours. Right after the curing compound, RP was covered using non-woven geotextile. The geotextile can absorb water. Cutting was done in the age range of 12- 18 hours according to project specifications. For this purpose, the geotextile is opened as needed. Therefore, about the first 2 hours, the RP has direct contact with the environment. The amounts of pores are largest until 2 hours [10]. After this time, moisture curing was done for seven days. The moisture-curing method is to pour water over the geotextile three times daily for seven days.

### *2.2. Methods*
The research method was done by observing defects per segment. Samples for bending and slump tests were taken from each concrete mixture from each dump truck. The slump test should be in the range of 0-5cm. Each segment condition was observed before the next step. Width and area of random defects, and also the first time the defects appear were noted. Three cores drill with 4" diameter done at the RP age about eight months. The core drill was done on the hair-crack area of the RP which was selected randomly. The depth of the cracks of every sample was measured. The depth of cracks determines the level of seriousness of the cracks for structural RP safety. The slump, grooving time, grooving period, curing compound quality, cutting period, surrounding weather condition are a parameter which was considered to analyze the cause of the random cracks. The parameters difference between no defect segment and segment with hair- cracks were studied. The early age behavior of the RP is assumed similar to [11] research. Based on the difference, and study literature, the cause of deep hair crack was analyzed.

## **3. Results and Discussion**
In this section we discuss how to format the title, authors and affiliations. Please follow these instructions as carefully as possible so all articles within a conference have the same style to the title page. This paragraph follows a section, title so it should not be indented.

### *3.1. Results*
Hair cracks in this RP were found in 231 segments at the age of 2 hours (figure 2). After the rain, hair-cracks appear clearer than before rain (fig. 2a and 2b). The width and depth of the hair-crack from core drill samples can be shown in fig. 2b and 2c. Slump at the segment with hair crack is closed to 5 cm (fig. 3).

Fine cracks appear on segments with high slumps. This causes grooving to be done later than the regular time so that the RP is longer exposed to the surrounding weather.

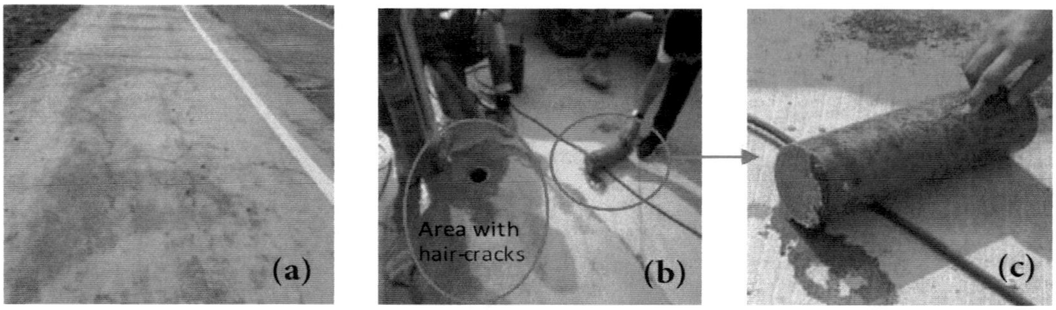


**Figure 2. a.** Hair-cracks after rainy, **b.** Hair-cracks before rain and core drilling RP with the hair crack, **c.** Sample 1 hair crack depth: 20 cm

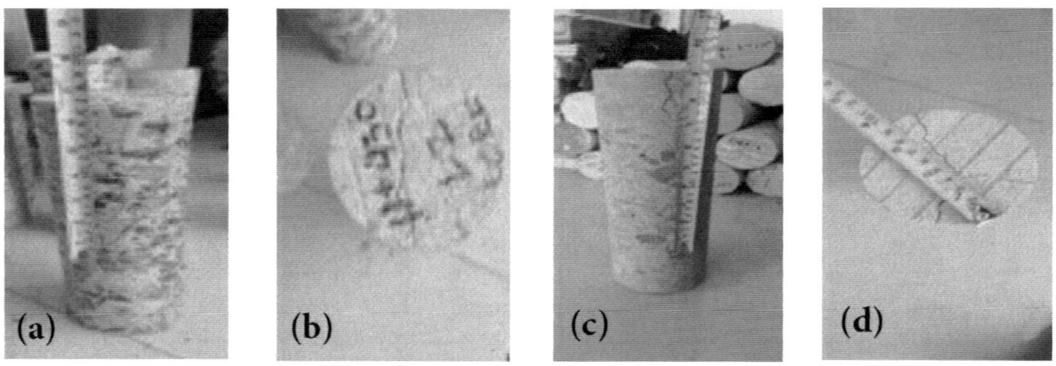


**Figure 3.** Sample 2 with hair-crack depth of 21 cm(a); hair-crack width of about 3mm (b, sample 3 with hair-crack depth of 23 cm (c), and hair-crack width of about 3mm

### *3.2. Discussion*

From observations and data obtained that the ratio between segments with hair-cracks and total segments is about 0.1%. Although it was small, another defects-type can be triggered by these cracks. These cracks influence surface defects index (SDI) value [12]. The crack depth ratio to the thickness of RP is in the range of 67-77%. Pouring concrete was done afternoon. High slump makes the setting time RP longer so that the waiting time for grooving is longer. This condition causes temperature, wind, and humidity to affect the moving out of inner water and makes RP surface more sensitive to cracks.

#### *3.2.1. Hydration*

Portland cement consisted of $C_3S$ (alite), $C_3A$, and $C_4AF$ (gypsum). $C_3S$ reaction is very fast resulted in CSH and CH. The expansion was caused by ettringite (Figure 4a) and $Ca^{2+}$ (Figure 4c). The $C_4AF$ was reacted to the CSH, and the product of this reaction is ettringite (Equation 1). The ettringite gels attach on the surface of cement grains until the concrete age of 18 hours [8] as displayed in Figure 4a.

$$C_4AF + 3CSH_2 + 30H \rightarrow C_6AS_3H_{32} + CH + FH_3 \quad (1)$$

At the age of 0-2 hours (hair cracks appear), ettringite or $C_6AS_3H_{32}$ (Figure 4a) and CH (Figure 4b) grow with a high rate for the first time, after that no growth of them both until the age of 6 hours [10]. The expansion of ettringite was offset by shrinkage because of CH. Because of this mechanism, the analysis of deep hair cracks mechanism was emphasized on $Ca^{2+}$ (Figure 4c). Portland cement contains 50%-80% of alite [13]; therefore, its product hydration dominates the behavior of concrete.

IOP Conf. Series: Materials Science and Engineering **857** (2020) 012028 doi:10.1088/1757-899X/857/1/012028

One of the product hydrations of alite is $Ca^{2+}$. Until the age of 1 hour, $Ca^{2+}$ growth at a high rate. The maximum number of alite was formed at the age of one hour.
After that, in the range age of 1-1.5 hours, the number of $Ca^{2+}$ decreases with a high rate [14]. Thus, there are two mechanisms during the first 2 hours, expansion and shrinkage. CH occupies about 15% of the volume of normal cement paste. It forms as crystals with a wide range of shapes and sizes, depending primarily on the amount of room available for growth [15].

En Wikipedia.org

(a) (b) (c)

**Figure 4. a.** Ettringite, **b.** CH (Portlandite), **c.** $Ca^{2+}$

*3.2.2. Environmental Effects*
The effect of temperature, ambient humidity, and rain on the appearance of hair cracks and deep hair crack is outlined as follow:

- *Hair cracks*
  Exposure to the environment makes the concrete more sensitive [16]. During the first 2 hours, this RP has directly contacted with surrounding weather. Ambient temperature is 27.3-28.7°C and humidity in the afternoon is 79-84%, respectively [11]. The autogenous concrete temperature in the age of 0-2 hours is between 27.6°C-29.1°C [11]. The difference of concrete temperature and ambient temperature is small (0.6°C); therefore, it can be assumed that no influences of ambient temperature to the concrete.

  A chemical reaction makes the concrete hot. A large amount of pore in the first 2 hours, makes surrounding relative humidity penetrate to the concrete. This condition makes CSH bonds were quicker formed.

  Because internal water was moving out caused by direct contact with the environment, hydration occurs with limited water. Water for hydration was obtained from pores water; therefore, on the surface, the tensile force occurs. Expansion and shrinkage due to $Ca^{2+}$ occur at a high rate. The change between expansion and decreasing volume needs bond strength and plastic. Both mechanisms happen together with moving out of internal water because of open conditions, lead tensile stress to be more. The high tensile stress breaks the weak bonds. The breaks make hair-cracks on the surface of RP. Beside CSH, the hydration was also produced CH. CH is the most soluble of the hydration products, and this is a weak link in cement and concrete [15-16].

- *Deep hair cracks*
  Deep hair-cracks were found from the core drill test at 8 months old. At the age, the hydration process is almost over. It is difficult to clearly distinguish between CH and CSH since the latter contains a significant proportion of Ca-OH bonds. CH contributes slightly to the strength and impermeability of the paste because it reduces the total pore volume by converting some of the liquid water into a solid form. Calcium hydroxide is believed to play

a role in limiting the amount of shrinkage that occurs when a cement paste is dried. As water is removed from the pore system, the CSH gel phase collapses, causing an overall shrinkage, while other crystalline phases such as CH are unaffected. As the CSH starts to shrink the CH that is in contact with it acts as a restraint so that the overall shrinkage is less than it would be if the CH were not present.

Blended cement pastes have little or no CH, as the CH that forms from cement hydration are consumed along with the mineral admixtures to form additional CSH gel. This condition has beneficial effects on the strength and permeability but tends to increase the tendency for drying shrinkage for the reasons mentioned above [15]. Freshwater from the surrounding (e.q. rainfall or dew) penetrates to the concrete from the hair-cracks. The freshwater, makes the CH leach out, increasing the porosity and makes the hair- cracks deeper and deeper. The phenomenon was making the paste more vulnerable to further leaching and chemical attack.

## 4. Conclusion

Tensile strength on the surface occurs because of removing water from the surface pores during two hours after casting. High rate expansion due to $Ca^{2+}$ makes the tensile force to be more. The shrinkage with a high rate occurs after that need strong and elastic chemical bonds. Removing water make CSH bonds which were still in weak condition collapse and can't handle the change from expansion to shrinkage and break. The breaks appear as hair-cracks. Penetration of freshwater (e.q. rainfall or dew) to the RP from the hair-cracks, removes the CH. With no CH in the concrete, the ability of concrete to handle deformation is decreasing This mechanism makes the increase of porosity, makes the cracks wider and deeper. It is very important to cover the rigid pavement right after pouring.

## Acknowledgment

This is to acknowledge the data and equipment supports from Pembangunan Perumahan Ltd for this research. Our appreciation is also goes to Universitas Lampung for supporting the research grant.

## References

[1] Ayers M, Cackler T, Fick G, Harrington D, Schwartz D, Smith K, Snyder M B, and Dam T V 2018 *Guide for Concrete Pavement Distress Assessment and Solution* (Iowa: National Concrete Pavement Technology Center) pp 1- 469

[2] Al-Kheetan M, Chamberlain D, and Rahman M 2019 *Int. J. Pavement Engineering* 1567917 pp **1-9**

[3] ACI 302 1RS 04 2006 *Guide for Concrete Floor and Slab Construction* (America: American Concrete Institute) pp 1-77

[4] Das U K, Bhardwaj B B, Choudhury B, and Sapkota G 2013 *IJAR* **3** 5 pp 222-223

[5] Transportation Research Circular 2006 *Control of Cracking in Concrete* E-C 107 Washington

[6] ACI 209R 92 1992 *Prediction of Creep, Shrinkage, and Temperature Effects in Concrete Structure* (America: American Concrete Institute) pp 1 – 47

[7] Hagelia P In: Mingshu T, and Min D 2004 12th International Conference on Alkali-Aggregate Reaction in Concrete (International Academic Publisher China: Beijing) 2 pp 870-881 Taylor H F W 1997 *Cement Chemistry* (London: Thomas Telford) p 459

[8] AASHTO M 148 2005 Standard Specification for Liquid Membrane-Forming Compounds for Curing Concrete pp 1-4

[9] Kurtis K 2015 *Portland Cement Hydration* (Georgia: Institute of Technology Atlanta) 35

[10] Niken C, Elly T, Supartono FX, Irianti L 2018 IOP Conf. Series: Materials Science and Engineering **316** 218) 012036

[11] Kementerian Pekerjaan Umum dan Perumahan Rakyat Direktorat Jendral Bina Marga 2016 *Integrated Road Management Systems Modul* (Indonesia)
[12] Kumar A, Bishnoi S, and Scrivener K L 2012 *Cem. Con. Res.* **42 903**-918
[13] Paulini P 1990 *J. Cem. Con. Res* **20 910**-918
[14] Thomas J, and Jennings H 2009 *The Science of Concrete* (USA: Infrastructure Technology Institute).
[15] Boualleg S, Bencheikh M, Belagraa L, Daoudi A, and Chikouche M A 2017 *J. Adv. Mater. Sci. Eng.* 2017 **5634713** p 17
[16] Richardson I, and Taylor H F W 2018 *Cement Chemistry* ICE Publishing pp 1-600.

# AUTHOR INDEX